D0628955

U.S. ARMY COMBAT PISTOL TRAINING MANUAL

DEPARTMENT OF THE ARMY

THE LYONS PRESS

Guilford, Connecticut
An imprint of The Globe Pequot Press

The Lyons Press is an imprint of The Globe Pequot Press.

10 9 8 7 6 5 4 3 2 1

Printed in the United States of America

Designed by Claire Zoghb

ISBN 1-59228-157-5

The Library of Congress Cataloging-in-Publication data is available on file.

CONTENTS

	Preface	vii
Chapter 1.	US Army Handguns	1
Chapter 2.	Marksmanship Training	7
	Section I. Basic Marksmanship	7
	Section II. Combat Marksmanship	21
	Section III. Coaching and Training Aids	31
	Section IV. Safety	41
Appendix A.	Combat Pistol Qualification Course	45
Appendix B.	Alternate Pistol Qualification Course	55
Appendix C.	Revolver Qualification Course	63
Appendix D.	Alternate Revolver Qualification Course	73
Appendix E.	Training Schedules	81
Appendix F.	Reproducible Forms	85
	Glossary	89
	References	91
	Index	95

U.S. ARMY COMBAT PISTOL TRAINING MANUAL

PREFACE

This manual provides guidance on the operation and marksmanship of the pistol, M9, 9-mm; pistol, M1911A1, caliber .45; and the revolver, caliber .38. It reflects current Army standards in weapons qualifications. It is a guide for the instructor to develop training programs, plans, and lessons that meet the objectives of the United States Army Marksmanship Program for developing combat effective marksmen. The soldier develops confidence, knowledge, and skills by following the guidelines in this manual.

The proponent of this publication is HQ TRADOC. Submit changes for improving this publication on DA Form 2028 (Recommended Changes to Publications and Blank Forms) and forward it to Commandant, U.S. Army Infantry School, ATTN: ATSH-IN-S3, Fort Benning, GA 31905-5593.

Unless otherwise stated, whenever the masculine gender is used, both men and women are included.

CHAPTER 1

US ARMY HANDGUNS

1-1. PISTOL, SEMIAUTOMATIC, 9-MM, M9

The M9 pistol is a 9-mm semiautomatic, magazine-fed, recoil-operated, double-action weapon chambered for the 9-mm cartridge. The magazine has a 15-round capacity.

a. Equipment Data

Caliber . . . . 9-mm NATO
System of Operation . . . . Short recoil, semiautomatic
Locking System . . . . Oscillating block
Length . . . . 217 mm (8.54 inches)
Width . . . . 38 mm (1.5 inches)
Height . . . . 140 mm (5.51 inches)
Weight with Empty Magazine . . . . 960 grams (2.1 pounds)
Weight with 15-Round Magazine . . . . . 1,145 grams (2.6 pounds)
Barrel Length . . . . 125 mm (4.92 inches)
Rifling . . . . . . R.H., six-groove (pitch 250 mm [about 10 inches])
Muzzle Velocity . . . . 375 mps (1,230.3 feet per second)
Muzzle Energy . . . . 569.5 newton meters (430 foot pounds)
Maximum Range . . . . 1,800 meters (1,962.2 yards)
Maximum Effective Range . . . . 50 meters (54.7 yards)
Front Sight . . . . Blade, integral with slide
Rear Sight . . . . Notched bar, dovetailed to slide
Sighting Radius . . . . 158 mm (6.22 inches)
Safety Features . . . . Decocking/safety lever, firing pin block
Hammer (half-cocked notch) . . . . Prevents accidental discharge
Basic Load . . . . 45 rounds

Trigger Pull. . . . . Single-Action: 5.50 pounds
Double-Action: 12.33 pounds

NOTE: For additional information on technical aspects of the M9 pistol see TM 9-1005-317-310.

★ ★ ★

WARNING

The half-cocked position catches the hammer and prevents it from firing if the hammer is released while manually cocking the weapon. It is not to be used as a safety position. The pistol will fire from the half-cocked position if the trigger is pulled.

★ ★ ★

b. Operation

The M9 pistol has a short recoil system using a falling locking block. The pressure developed by the expanding gases of a fired round recoils the slide and barrel assembly. After a short run, the locking block is disengaged from the slide, the barrel stops against the frame, and the slide continues its rearward movement. The slide then extracts and ejects the fired cartridge case, cocks the hammer, and compresses the recoil spring. The slide moves forward feeding the cartridge from the magazine into the chamber. The slide and barrel assembly remain open after the last cartridge has been fired and ejected.

1-2. PISTOL, AUTOMATIC, .45 CALIBER, M1911 AND M1911A1

The M1911 and M1911A1 pistols are semiautomatic, .45-caliber, recoil-operated, magazine-fed, single-action pistols. The magazine has a seven-round capacity.

a. Equipment Data

Caliber. . . . . 0.45 inches
System of Operation. . . . . Short recoil, semiautomatic

Length . . . . 8 5/8 inches
Weight with Empty Magazine . . . . 2.4 pounds
Weight with Full Magazine. . . . . 3 pounds
Length of Barrel . . . . 5.03 inches
Rifling. . . . . L.H., six-groove (Pitch 1 in 16 inches)
Muzzle Velocity. . . . . 830 feet per second
Muzzle Energy . . . . 17,000 pounds per square inch
Maximum Range. . . . . 1,500 meters
Maximum Effective Range . . . . 50 meters
Front Sight. . . . . Blade, integral with slide
Rear Sight. . . . . Notched bar, dovetailed to slide
Sight Radius . . . . 6.481 inches
Safety . . . . . . Manual safety lever, grip safety, half-cock position
Basic Load . . . . 21 rounds
Trigger Pull. . . . . 5 to 6 1/2 pounds

b. Operation

(1) Each time a cartridge is fired, the parts inside the weapon function in a given order. This is known as the *functioning cycle* or *cycle of operation.*

(2) The cycle of operation of the weapon is divided into eight steps: feeding, cambering, locking, firing, unlocking, extracting, ejecting, and cocking. The steps are listed in the order in which functioning occurs; however, more than one step may occur at the same time.

(3) A magazine containing ammunition is placed in the receiver. The slide is pulled fully to the rear and released. As the slide moves forward, it strips the top round from the magazine and pushes it into the chamber. The hammer remains in the cocked position, and the weapon is ready to fire.

(4) The weapon fires one round each time the trigger is pulled. Each time a cartridge is fired, the slide and barrel recoil or move a short distance locked together. This permits the bullet and expanding powder gases to escape from the muzzle before the unlocking is completed.

(5) The barrel then unlocks from the slide and continues to

the rear, extracting the cartridge case from the chamber and ejecting it from the weapon. During this rearward movement the magazine feeds another cartridge, the recoil spring is compressed, and the hammer is cocked.

(6) At the end of the rearward movement, the recoil spring expands, forcing the slide forward, locking the barrel and slide together. The weapon is ready to fire again. The same cycle of operation continues until the ammunition is expended.

(7) As the last round is fired, the magazine spring exerts upward pressure on the magazine follower. The stop on the follower strikes the slide stop, forcing it into the recess on the bottom of the slide and locking the slide to the rear. This action indicates that the magazine is empty and aids in faster reloading.

NOTE: For additional information on the technical aspects of the caliber .45 pistol see TM 9-1005-211-12.

1-3. REVOLVER, CALIBER .38

There are six basic caliber. 38 service revolvers in use by the Army. One is a 2-inch barreled, .38-caliber revolver made by Smith and Wesson; five are 4-inch barreled, .38-caliber revolvers—three made by Ruger, and two by Smith and Wesson. The 2-inch barreled revolver is used mainly by Army CID and counterintelligence personnel. The 4-inch barreled revolvers are used by aviators and military police.

a. Equipment Data

Smith and Wesson

Caliber. . . . . 0.38 inches
System of Operation. . . . . Rotated chamber
Length: 2-Inch Barrel. . . . . 7 1/4 inches
4-Inch Barrel. . . . . 9 1/4 inches
Weight: 2-Inch Barrel . . . . 26.5 ounces
4-Inch Barrel . . . . 30.5 ounces
Length of Barrel. . . . . 2 inches/4 inches
Muzzle Velocity. . . . . 950 feet per second
Muzzle Energy . . . . 16,000 per square inch

Maximum Range: 2-Inch Barrel .... 868 meters
4-Inch Barrel .... 992 meters
Maximum Effective Range .... 45 meters (2-inch barrel)
60 meters (4-inch barrel)
Front Sight .... Fixed 1/8-inch serrated ramp
Rear Sight .... Square notch
Safety Features .... No manually operated safety
Basic Load .... 18 rounds

Ruger

Caliber .... 0.38 inches
System of Operation .... Rotated chamber
Length .... 9 1/4 inches
Weight .... 33 ounces
Length of Barrel .... 4 inches
Muzzle Velocity .... 950 feet per second
Muzzle Energy .... 16,000 per square inch
Maximum Range .... 992 meters
Maximum Effective Range .... 60 meters
Front Sight .... Fixed blade
Rear Sight .... Fixed groove
Safety Features .... No manually operated safety
Basic Load .... 18 rounds

b. Operation.

(1) When firing single-action, the hammer is pulled back, and the sear engages the full-cock notch in the hammer.

(a) Smith and Wesson: Pulling the trigger lowers the hammer block, allowing the hammer to fall.

(b) Ruger: Pulling the trigger raises the transfer bar into the firing position between the hammer and firing pin, allowing the hammer to strike the firing pin.

(2) When firing double-action, the trigger is squeezed. This engages the sear, raising the hammer to nearly full-cock position. Continued pressure on the trigger allows the sear to escape from the trigger and the hammer to fall.

(a) Smith and Wesson: When the trigger is squeezed, the rebound slide pivots the hammer block downward, striking the cartridge primer.

(b) Ruger: When the trigger is squeezed and held to the rear, the transfer bar passes force from the transfer bar to the firing pin, striking the cartridge primer. If the trigger is not held to the rear, the hammer rests directly on the frame and the transfer bar remains below the firing pin.

(3) The cylinder stop (Smith and Wesson) or latch (Ruger) prevents the cylinder from making more than one-sixth of a revolution each time the weapon is cocked. The cylinder stop/latch withdraws from the cylinder as the trigger moves. The trigger hand (Smith and Wesson) or pawl (Ruger) pivots and engages the ratchet on the extractor/ejector portion of the cylinder. The trigger slips off of the cylinder stop/latch as it continues rearward. The cylinder stop/latch then engages the next notch.

NOTES: 1. In firing the Ruger, the trigger must remain all the way back till the hammer falls. If the trigger is released before the hammer falls, the weapon will not fire. In firing the Smith and Wesson, the weapon fires only when the trigger is pulled all the way back.

2. For additional information on the technical aspects of the caliber .38 see TM 9-1005-226-14 and TM 9-1005-205-14&P-1.

CHAPTER 2

MARKSMANSHIP TRAINING

SECTION I.
BASIC MARKSMANSHIP

2-1. PHASES OF TRAINING

Marksmanship training is divided into two phases: preparatory marksmanship training and range firing. Each phase may be divided into separate instructional steps. All marksmanship training must be progressive. Combat marksmanship techniques should be practiced after the basics have been mastered.

2-2. FUNDAMENTALS

The main use of the pistol or revolver is to engage an enemy at close range with quick, accurate fire. Accurate shooting results from knowing and correctly applying the elements of marksmanship. The elements of combat pistol or revolver marksmanship are:

- Grip.
- Aiming.
- Breath control.
- Trigger squeeze.
- Target engagement.
- Positions.

2-3. GRIP

The weapon must become an extension of the hand and arm. It should replace the finger in pointing at an object. A firm, uniform grip must be applied to the weapon. A proper grip is one of the most important fundamentals of quick fire.

a. One-Hand Grip. Hold the weapon in the nonfiring hand; form a V with the thumb and forefinger of the strong hand (firing hand) (see Figure 2-1). Place the weapon in the V with the front and rear sights in line with the firing arm. Wrap the lower three fingers around the pistol grip, putting equal pressure with all three fingers to the rear. Allow the thumb of the firing hand to rest alongside the weapon *without* pressure. Grip the weapon tightly until the hand begins to tremble; relax until the trembling stops. At this point, the necessary pressure for a proper grip has been applied. Place the trigger finger on the trigger between the tip and second joint so that it can be squeezed to the rear. The trigger finger must work independently of the remaining fingers.

NOTE: If any of the three fingers on the grip is relaxed the grip must be reapplied.

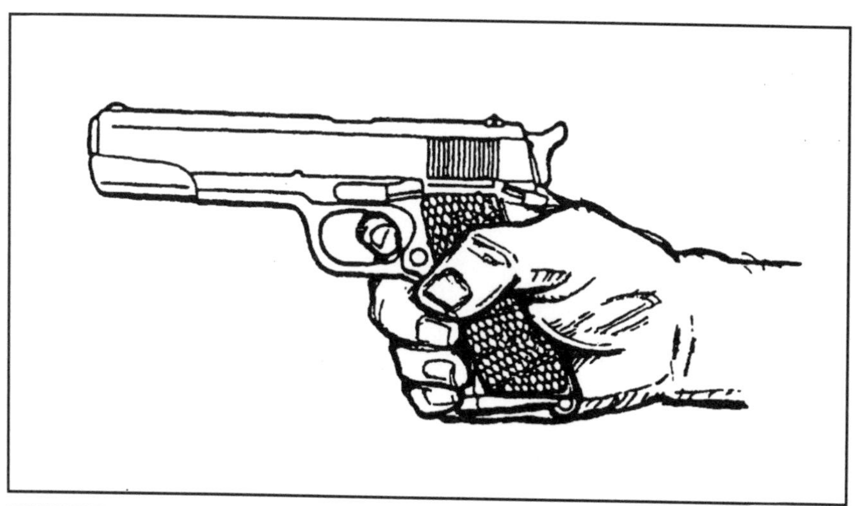

FIGURE 2-1. One-hand grip.

b. Two-Hand Grip. The two-hand grip allows the firer to steady the firing hand and provide maximum support during firing. The nonfiring hand becomes a support mechanism for the firing hand by wrapping the fingers of the nonfiring hand around the firing hand. Two-hand grips are recommended for all pistol and revolver firing.

★ ★ ★

WARNING

If the nonfiring thumb is placed in the rear of the weapon, the recoil from the weapon could result in personal injury.

★ ★ ★

(1) *Fist grip.* Grip the weapon as described in **paragraph a** above. Firmly close the fingers of the nonfiring hand over the fingers of the firing hand, ensuring that the index finger from the nonfiring hand is between the middle finger of the firing hand and the trigger guard. Place the nonfiring thumb alongside the firing thumb. (See Figure 2-2.)

NOTE: Depending upon the individual firer, he may choose to place his index finger of the nonfiring hand on the front of the trigger guard of the M9 pistol since this weapon has a recurved trigger guard designed for this purpose.

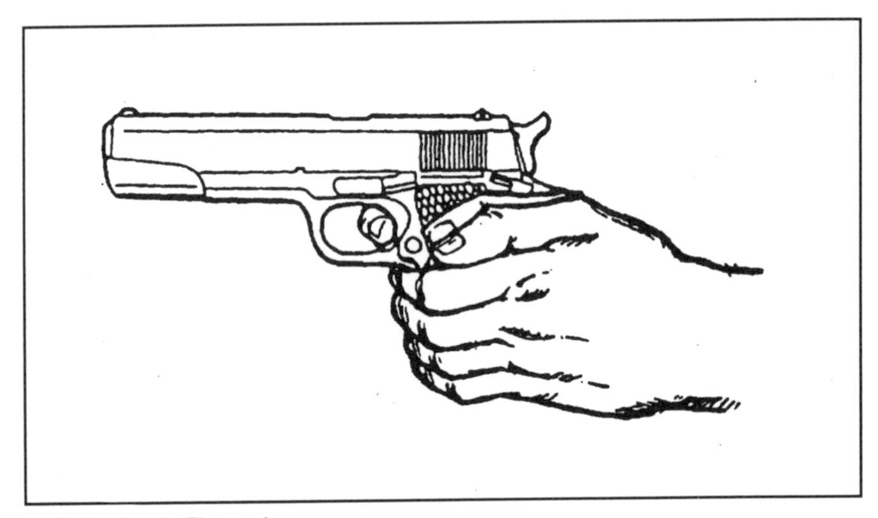

FIGURE 2-2. First grip.

(2) *Palm-supported grip.* This grip is commonly called the cup and saucer grip. Grip the firing hand as described in **paragraph a**

above. Place the nonfiring hand under the firing hand, wrapping the nonfiring fingers around the back of the firing hand. Place the nonfiring thumb over the middle finger of the firing hand. (See Figure 2-3.)

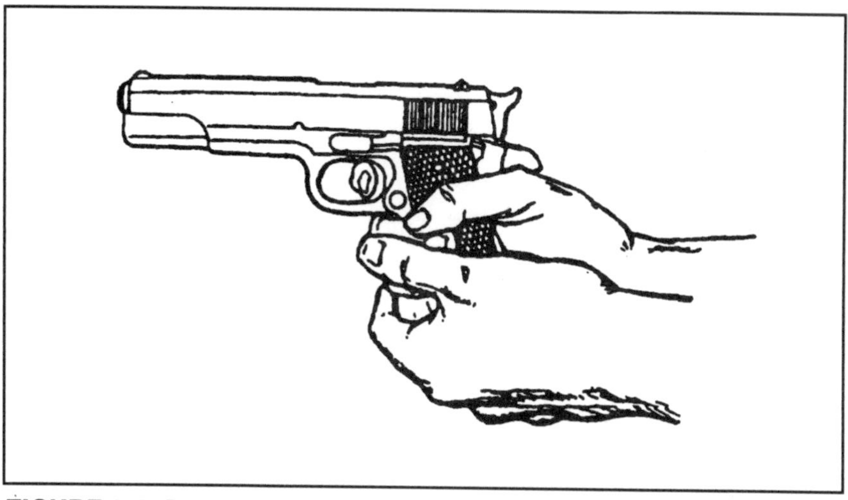

FIGURE 2-3. Palm-supported grip.

(3) *Weaver grip.* Apply this grip the same as the fist grip. The only exception is that the nonfiring thumb is wrapped over the firing thumb. (See Figure 2-4.)

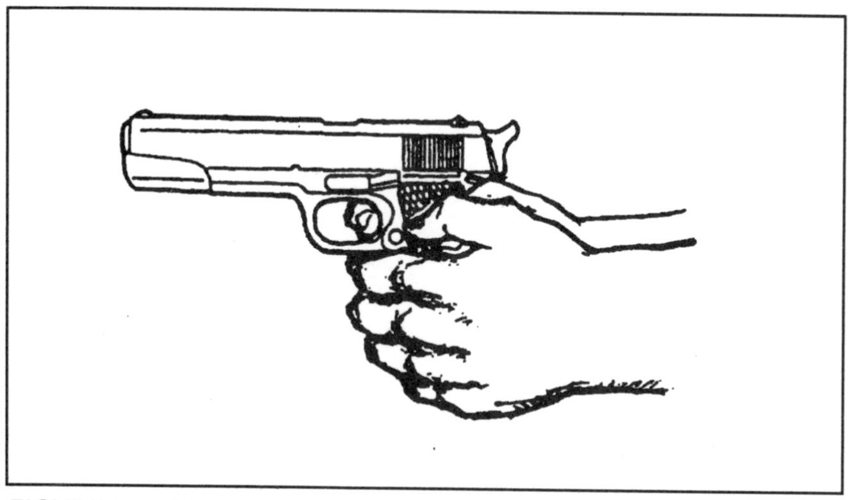

FIGURE 2-4. Weaver grip.

c. Isometric Tension. The firer raises his arms to a firing position and applies isometric tension. This is commonly known as the push-pull method for maintaining weapon stability. Isometric tension is when the firer applies forward pressure with the firing hand and pulls rearward with the nonfiring hand with equal pressure. This creates an isometric force but never so much to cause the firer to tremble. This steadies the weapon and reduces barrel rise from recoil. The supporting arm is bent with the elbow pulled downward. The firing arm is fully extended with the elbow and wrist locked. The firer must experiment to find the right amount of isometric tension to apply.

NOTE: The firing hand should exert the same pressure as the nonfiring hand. If it does not, a missed target could result.

d. Natural Point of Aim. The firer should check his grip for use of his natural point of aim. He grips the weapon and sights properly on a distant target. While maintaining his grip and stance, he closes his eyes for three to five seconds. He then opens his eyes and checks for proper sight picture. If the point of aim is disturbed, the firer adjusts his stance to compensate. If the sight alignment is disturbed, the firer adjusts his grip to compensate by removing the weapon from his hand and reapplying the grip. The firer repeats this process until the sight alignment and sight placement remain almost the same when he opens his eyes. This enables the firer to determine and use his natural point of aim once he has sufficiently practiced. This is the most relaxed position for holding and firing the weapon.

2-4. AIMING

a. Aiming is sight alignment and sight placement (see Figure 2-5). Sight alignment is the centering of the front blade in the rear sight notch. The top of the front sight is level with the top of the rear sight and is in correct alignment with the eye. For correct sight alignment, the firer must center the front sight in the rear sight. He raises or lowers the top of the front sight so it is level with the top of the rear sight.

b. Sight placement is the positioning of the weapon's sights in relation to the target as seen by the firer when he aims the weapon (see

Figure 2-5). A correct sight picture consists of correct sight alignment with the front sight placed center mass of the target. The eye can focus on only one object at a time at different distances. Therefore the last focus of the eye is always on the front sight. When the front sight is seen clearly, the rear sight and target will appear hazy. Correct sight alignment can only be maintained through focusing on the front sight. The firer's bullet will hit the target even if the sight picture is partly off center but still remains on the target. Therefore, sight alignment is more important than sight placement. Since it is impossible to hold the weapon completely still, the firer must apply trigger squeeze and maintain correct sight alignment while the weapon is moving in and around the center of the target. This natural movement of the weapon is referred to as *wobble area*. The firer must strive to control the limits of the wobble area through proper breath control, trigger squeeze, positioning, and grip.

c. Sight alignment is essential for accuracy because of the short sight radius of the pistols and revolvers. For example, if a 1/10-inch error is made in aligning the front sight in the rear sight, the firer's bullet will miss the point of aim by about 15 inches at a range of 25 meters. The 1/10-inch error in sight alignment magnifies as the range increases—at 25 meters it is magnified 150 times.

d. Focusing on the front sight while applying proper trigger squeeze will help the firer resist the urge to jerk the trigger and anticipate the actual moment the weapon will fire. Mastery of trigger

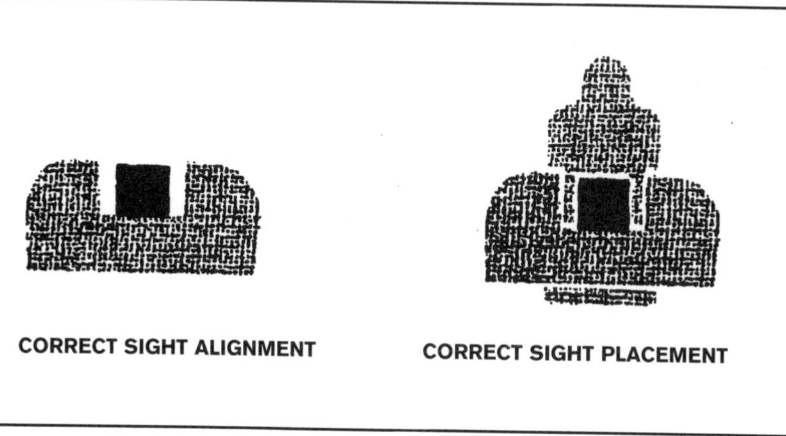

FIGURE 2-5. Correct sight alignment and sight picture.

squeeze and sight alignment requires practice. Trainers should use concurrent training stations or have fire ranges to enhance proficiency of marksmanship skills.

2-5. BREATH CONTROL

The firer must learn to hold his breath properly at any time during the breathing cycle if he wishes to attain accuracy that will serve him in

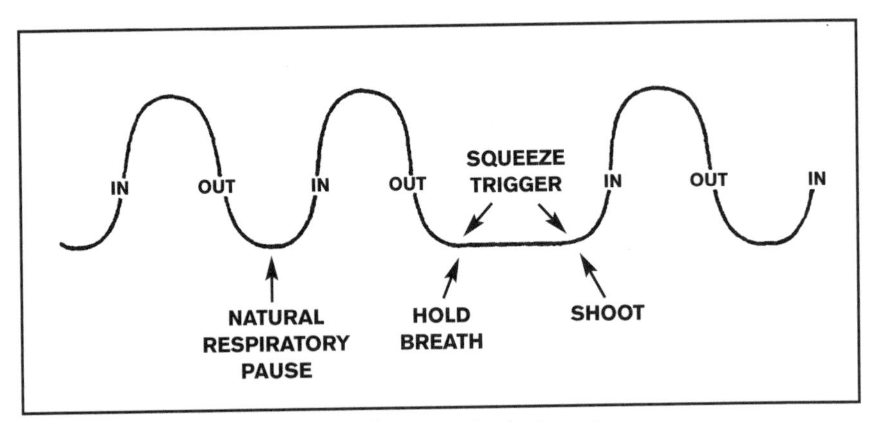

FIGURE 2-6. Breath control, firing at a single target.

combat. This must be done while aiming and squeezing the trigger. While the procedure is simple, it requires explanation, demonstration, and supervised practice. To hold the breath properly, the firer takes a breath, lets it out, then inhales normally, lets a little out until comfortable, holds, and then fires. It is difficult to maintain a steady position

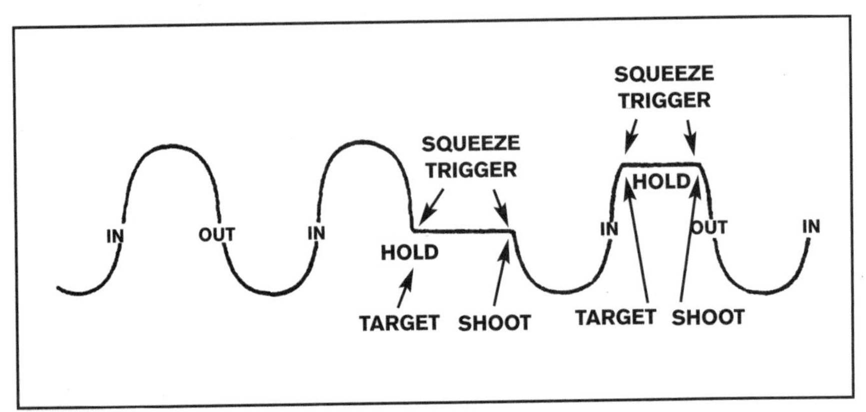

FIGURE 2-7. Breath control, firing at a timed or multiple targets.

keeping the front sight at a precise aiming point while breathing. Therefore, the firer should be taught to inhale, then exhale normally, and hold his breath at the moment of the natural respiratory pause (see Figure 2-6). The shot must then be fired before he feels any discomfort from not breathing. When multiple targets are presented, the firer must learn to hold his breath at any part of the breathing cycle (see Figure 2-7). Breath control must be practiced during dry-fire exercises until it becomes a natural part of the firing process.

2-6. TRIGGER SQUEEZE

a. Improper trigger squeeze causes more misses than any other step of preparatory marksmanship. Poor shooting is caused by the aim being disturbed before the bullet leaves the barrel of the weapon. This is usually the result of the firer jerking the trigger or flinching. A slight off-center pressure of the trigger finger on the trigger can cause the weapon to move and disturb the firer's sight alignment. Flinching is an automatic human reflex caused by anticipating the recoil of the weapon. Jerking is an effort to fire the weapon at the precise time the sights align with the target.

NOTE: See problems in target engagement, paragraph 2-7.

b. Trigger squeeze is the independent movement of the trigger finger in applying increasing pressure on the trigger straight to the rear, without disturbing the sight alignment until the weapon fires. The trigger slack, or free play, is taken up first, and the squeeze is continued steadily until the hammer falls. If the trigger is squeezed properly, the firer will not know exactly when the hammer will fall; thus, he does not tend to flinch or heel, resulting in a bad shot. Novice firers must be trained to overcome the urge to anticipate recoil. Proper application of the fundamentals will lower this tendency.

c. To apply correct trigger squeeze, the trigger finger should contact the trigger between the tip of the finger and the second joint (without touching the weapon anywhere else). Where contact is made depends on the length of the firer's trigger finger. If pressure from the trigger finger is applied to the right side of the trigger or weapon, the strike of the bullet will be to the left. This is due to the normal hinge action of the fingers. When the fingers on the right hand are closed, as in gripping, they hinge or pivot to the left, thereby applying pressure to the left. (With left-handed firers, this action is to the right.) The firer must not apply pressure left or right but increase finger pressure straight to the rear. Only the trigger finger must perform this action. Dry-fire training improves a firer's ability to move the trigger finger straight to the rear without cramping or increasing pressure on the hand grip.

(1) The firer who is a good shot holds the sights of the weapon as nearly on the target center as possible and continues to squeeze the trigger with increasing pressure until the weapon fires.

(2) The soldier who is a bad shot tries to "catch his target" as his sight alignment moves past the target and fires the weapon at that instant. This is called *ambushing*, which causes trigger jerk.

d. Follow-through is the continued effort of the firer to maintain sight alignment before, during, and after the round has fired. The firer must continue the rearward movement of the finger even after the round has been fired. Releasing the trigger too soon after the round has been fired results in an uncontrolled shot, causing a missed target.

NOTE: The trigger squeeze of the M9 pistol, when fired in the single-action mode, is 5.50 pounds; when fired in double-action mode, it is 12.33 pounds. The firer must be aware of the mode he is firing in. He must also practice squeezing the trigger in each mode to develop expertise in single-action and double-action target engagements.

2-7. TARGET ENGAGEMENT

To engage a single target, the firer applies the method discussed in paragraph 2-6 when multiple targets are engaged. The closest and most dangerous multiple target in combat is engaged first and should be fired at with two rounds. This is commonly referred to as a *double tap*. The firer then traverses and acquires the next target, aligns the sights in the center of mass, focuses on the front sight, applies trigger squeeze, and fires. The firer ensures his firing arm elbow and wrist are locked during all engagements. If the firer has missed the first target and has fired upon the second target, he shifts back to the first and engages it. Some problems in target engagement are as follows:

a. Recoil Anticipation. When a soldier first learns to shoot, he may begin to anticipate recoil. This reaction may cause him to tighten his muscles during or just before the hammer falls. He may fight the recoil by pushing the weapon downward in anticipating or reacting to its firing. In either case, the rounds will not hit the point of aim. A good method to show the firer that he is anticipating the recoil is the ball-and-dummy method (see paragraph 2-16).

b. Trigger Jerk. Trigger jerk occurs when the soldier sees that he has acquired a good sight picture at center mass and "snaps" off a round before the good sight picture is lost. This may become a problem, especially when the soldier is learning to use a flash sight picture (see paragraph 2-9).

c. Heeling. Heeling is caused by a firer tightening the large muscle in the heel of the hand to keep from jerking the trigger. A firer who has had problems with jerking the trigger tries to correct the fault by tightening the bottom of the hand, which results in a heeled shot. Heeling causes the strike of the bullet to hit high on the firing hand side of the target. The firer can correct shooting errors by knowing and applying correct trigger squeeze.

2-8. POSITIONS

The qualification course is fired from a standing, kneeling, or crouch position. All of the firing positions described below must be practiced so they become natural movements, during qualification and combat firing. Though these positions seem natural, practice sessions must be conducted to ensure the habitual attainment of correct firing positions. Assuming correct firing positions ensures that soldiers can quickly assume these positions without a conscious effort. Pistol marksmanship requires a soldier to rapidly apply all the fundamentals at dangerously close targets while under stress. Assuming a proper position to allow for a steady aim is critical to survival.

a. Pistol-Ready Position. In the pistol-ready position, hold the weapon in the one-hand grip. Hold the upper arm close to the body,

• STEP 4: Know which reloading procedure to use for the tactical situation. There are three systems of reloading: rapid, tactical, and one-handed. *Rapid reloading* is used when the soldier's life is in immediate danger, and the reload must be accomplished quickly. *Tactical reloading* is used when there is more time, and it is desirable to keep the replaced magazine because there are rounds still in it or it will be needed again. *One-handed reloading* is used when there is an arm injury.

a. Rapid Reloading.

• Place your hand on the next magazine in the ammunition pouch to ensure there is another magazine.

• Withdraw the magazine from the pouch while releasing the other magazine from the weapon. Let the replaced magazine drop to the ground.

• Insert the replacement magazine, guiding it into the magazine well with the index finger.

• Release the slide, if necessary.

• Pick up the dropped magazine if time allows. Place it in your pocket, not back into the ammunition pouch where it may become mixed with full magazines.

b. Tactical Reloading.

• Place your hand on the next magazine in the ammunition pouch to ensure there is a remaining magazine.

• Withdraw the magazine from the pouch.

• Drop the used magazine into the palm of the nonfiring hand, which is the same hand holding the replacement magazine.

• Insert the replacement magazine, guiding it into the magazine well with the index finger.

• Release the slide, if necessary.

• Place the used magazine into a pocket. Do not mix it with full magazines.

c. One-Handed Reloading.

(1) With the right hand.

• Push the magazine release button with the thumb.

• Place the safety ON with the thumb if the slide is forward.

• Place the weapon backwards into the holster.

NOTE: If placing the weapon in the holster backwards is a problem, place the weapon between the calf and thigh to hold the weapon.

- Insert the replacement magazine.
- Withdraw the weapon from the holster.
- Remove the safety with the thumb if the slide is forward, or push the slide release if the slide is back.

(2) With the left hand.

- Push the magazine release button with the middle finger.
- Place the safety ON with the thumb if the slide is forward. With the .45-caliber pistol, the thumb must be switched to the left side of the weapon.
- Place the weapon backwards into the holster.

NOTE: If placing the weapon in the holster backwards is a problem, place the weapon between the calf and thigh to hold the weapon.

- Insert the replacement magazine.
- Remove the weapon from the holster.
- Remove the safety with the thumb if the slide is forward, or push the slide release lever with the middle finger if the slide is back.

2-13. POOR VISIBILITY FIRING

Poor visibility firing with any weapon is difficult since shadows can be misleading to the soldier. This is mainly true during EENT and EMNT (a half hour before dark and a half hour before dawn). Even though the weapon is a short-range weapon, the hours of darkness and poor visibility further decrease its effect. To compensate, the soldier must use the three principles of night vision.

a. Dark Adaptation. This process conditions the eyes to see during poor visibility conditions. The eyes usually need about 30 minutes to become 98-percent dark adapted in a totally darkened area.

b. Off-Center Vision. When looking at an object in daylight, a person looks directly at it. However, at night he would see the object only for a few seconds. To see an object in darkness, he must concentrate on it while looking 6° to 10° away from it.

c. Scanning. This is the short, abrupt, irregular movement of the firer's eyes around an object or area every 4 to 10 seconds. When artificial illumination is used, the firer uses night fire techniques to engage targets, since targets seem to shift without moving.

NOTE: For more detailed information on the three principles of night vision, see FM 21-75.

2-14. NUCLEAR, BIOLOGICAL, CHEMICAL FIRING

When firing under NBC conditions with a pistol or revolver, the firer should use optical inserts, if applicable. Firing in MOPP1 through MOPP3 levels should not be a problem for the firer. Unlike wearing a protective mask while firing a rifle, the firer's sight picture will be acquired the same as with or without a protective mask. MOPP4 is the only level that may present a problem for a firer since gloves are worn. Gloves may require the firer to adjust his grip to attain a proper grip and proper trigger squeeze. Firers should practice firing in MOPP4 to become proficient in NBC firing.

SECTION III.
COACHING AND TRAINING AIDS

2-15. COACHING

a. Throughout preparatory marksmanship training, the coach-and-pupil method of training should be used. The proficiency of a pupil depends on how well his coach performs his duties. The coach assists the firer by correcting errors, ensuring he takes proper firing positions, and ensuring he observes all safety precautions. The criteria for selecting coaches are a command responsibility; coaches must have experience in pistol marksmanship above that of the student firer.

b. Duties of the coach during instruction practice and record firing include:

(1) Checking that the–

(a) Weapon is cleared.

(b) Ammunition is clean.

(c) Magazines are clean and operational.

(2) Observing the firer to see that he–

(a) Takes the correct firing position.

(b) Loads the weapon properly and only on command.

(c) Takes up the trigger slack correctly.

(d) Squeezes the trigger correctly (see paragraph 2-6).

(e) Calls the shot each time he fires (except for quick fire and rapid fire).

(f) Holds his breath correctly (see paragraph 2-5).

(g) Lowers his weapon and rests his arm when he does not fire a round within five to six seconds.

(3) Having the firer breathe deeply several times to relax if he is tense.

2-16. BALL-AND-DUMMY METHOD

In this method the coach loads the weapon for the firer. He may hand the firer a loaded weapon or an empty one. When firing the empty weapon, the firer observes that in anticipating recoil he is forcing the weapon downward as the hammer falls. Repetition of the ball-and-dummy method helps to alleviate recoil anticipation.

2-17. CALLING THE SHOT

To call the shot is to state where the bullet should strike the target according to the sight picture at the instant the weapon fires—for example: "high," "a little low," "to the left," "to the right," or "bull's-eye." If the firer does not call his shot correctly in range firing, he is not concentrating on sight alignment. Consequently, he does not know what his sight picture is as he fires. Another method of calling the shot is the clock system—for example, a three-ring hit at 8 o'clock, a four-ring hit at 3 o'clock. Another method is to provide the firer with a target center (placed beside him on the firing line). As soon as the shot is fired, the firer must place a finger on the target face or center where he expected the round to hit the target. This method avoids guessing and computing for the firer. The immediate placing of the finger on the target face gives an accurate call. If the firer does not call his shot correctly, he is not concentrating on sight alignment and trigger squeeze. Thus, he does not know what his sight picture is as the weapon fires.

2-18. PENCIL TRIANGULATION EXERCISE

The pencil triangulation exercise (see Figure 2-25) is conducted only with an unloaded and properly cleared M1911A1 caliber .45 pistol. It will not work with an M9 pistol; however, a coach may have his students dry fire the M9 while he observes the firers to see if the front sight dips or jumps when the hammer falls. The pencil triangulation exercise consists of firing a pencil or pointed dowel point-blank at a miniature target. It combines position, grip, sight alignment, breathing, and trigger squeeze into a single practical work exercise. At the same time, it measures the firer's performance without the effects of recoil. This practical work is designed to teach and develop correct shooting

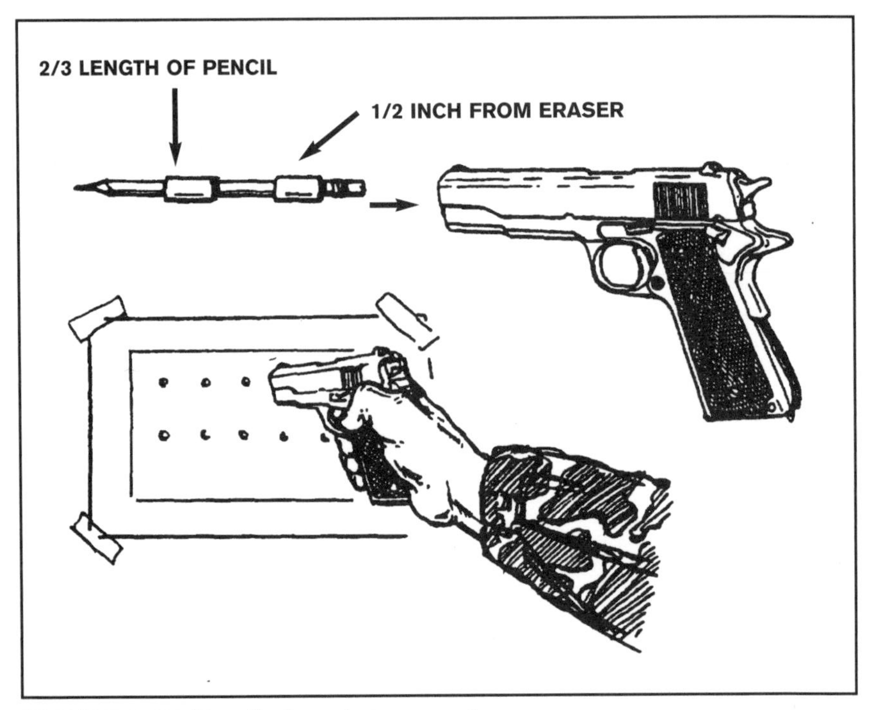

FIGURE 2-25. Pencil triangulation exercise.

habits. It can be conducted indoors or out, which makes an ideal exercise where range facilities are limited or when weather is poor.

a. Equipment

(1) One dowel or lead pencil for every two students. This pencil should be at least 6 inches long and wrapped with masking or cellophane tape. The tape wrappings form two bushings that fit the inside diameter of the weapon's barrel.

(2) One miniature bull's-eye sheet for every two students. The bull's-eye sheet can be copied, drawn, or stamped by using the eraser of a pencil and ink pad. The bull's-eyes should not be larger than 1/8 inch and at least 1 inch apart.

b. Conduct of the Exercise. The instructor explains and demonstrates the details of the exercise before practical work by the students. The firer should begin by using a two-hand grip, progressing to the one-hand grip as his skills increase.

(1) The firer faces the target and takes up a good shooting position. This position is close enough to the miniature bull's-eye that when the pencil is inserted into the barrel, with the firer's arm extended and the sights aimed at the miniature bull's-eye, the point of the pencil is within 1 inch of the target. The bull's-eye sheet should be affixed to a target, or any type support, and should be shoulder-high to the firer.

(2) The firer inserts the pencil into the muzzle of the barrel, eraser end first, and cocks the hammer. He grips the weapon properly, extends the shooting arm, aims the weapon at the miniature bull's-eye, squeezes the trigger, and the hammer falls. The hammer strikes the firing pin, which in turn strikes the rubber eraser of the pencil, driving it out of the barrel and causing it to make a pencil dot 1/2 inch below the bull's-eye (if the firer had the correct sight alignment and trigger squeeze).

(3) The firer continues this exercise until he has fired a group of five pencil marks below each target. The object of the exercise is to keep the five pencil marks in a group as small as the 1/8-inch bull's-eye, 1/2 inch directly below the bull's-eye. With practice, many firers can hit the same mark with the pencil. This indicates that the firer is properly performing the fundamentals of marksmanship each time.

2-19. SLOW-FIRE EXERCISE

a. This is a dry-fire exercise. The slow-fire exercise is one of the most important exercises for both amateur and competitive marksmen. Coaches should ensure soldiers practice this exercise as much as possible. To perform the slow-fire exercise, the firer assumes the standing position with the weapon pointed at the target. The firer should begin by using a two-hand grip, progressing to the one-hand grip as his skill increases. He takes in a normal breath and lets part of it out, locking the remainder in his lungs by closing his throat. He then relaxes, aims at the target, takes the correct sight alignment and sight picture, takes up the trigger slack, and squeezes the trigger straight to the rear with steady, increasing pressure until the hammer falls, simulating firing.

b. If the firer does not cause the hammer to fall in 5 or 6 seconds, he should come to the pistol-ready position, and rest his arm and hand. He then starts the procedure again. The action sequence that makes up this process can be summed up by the key word BRASS. It is a word the firer should think of each time he fires his weapon:

Breathe—Take a normal breath, let part of it out, and lock the remainder in the lungs by closing the throat.

Relax—Relax the body muscles.

Aim—Take correct sight alignment and sight picture, and focus the eye at the top of the front sight.

Slack—Take up the trigger slack.

Squeeze—Squeeze the trigger straight to the rear with steadily increasing pressure without disturbing sight alignment until the hammer falls.

c. Coaches should observe the front sight for erratic movements during the application of trigger squeeze. Proper application of trigger squeeze allows the hammer to fall without the front sight moving. A small bouncing movement of the front sight is acceptable. Firers should call the shot by the direction of movement of the front sight (high, low, left, or right).

2-20. AIR-OPERATED PISTOL, .177 MM

The air-operated pistol is used as a training device to teach the soldier the method of quick fire, to increase his confidence in his ability, and to afford him more practice firing. A range can be set up almost anywhere with a minimum of effort and coordination, which is ideal for USAR and NG. If conducted on a standard range, live firing of pistols and revolvers can be conducted along with the firing of the .177-mm air-operated pistol. Due to the light recoil and little noise of the pistol, the soldier can concentrate on fundamentals. This helps build confidence, because the soldier can hit a target faster and more accurately. The air-operated pistol should receive the same respect as any firearm. A thorough explanation of the weapon and a safety briefing are given to each soldier.

2-21. QUICK-FIRE TARGET TRAINING DEVICE

The QTTD (see Figures 2-26 and 2-27) is used with the .177-mm air-operated pistol.

PHASE I. From 10 feet, five shots at a 20-foot miniature E-type silhouette. After firing each shot, the firer and coach discuss the results and make corrections.

PHASE II. From 15 feet, five shots at a 20-foot miniature E-type silhouette. The same instructions apply to this exercise as for PHASE I.

PHASE III. From 20 feet, five shots at a 20-foot miniature E-type silhouette. The same instructions apply to this exercise as for PHASES I and II.

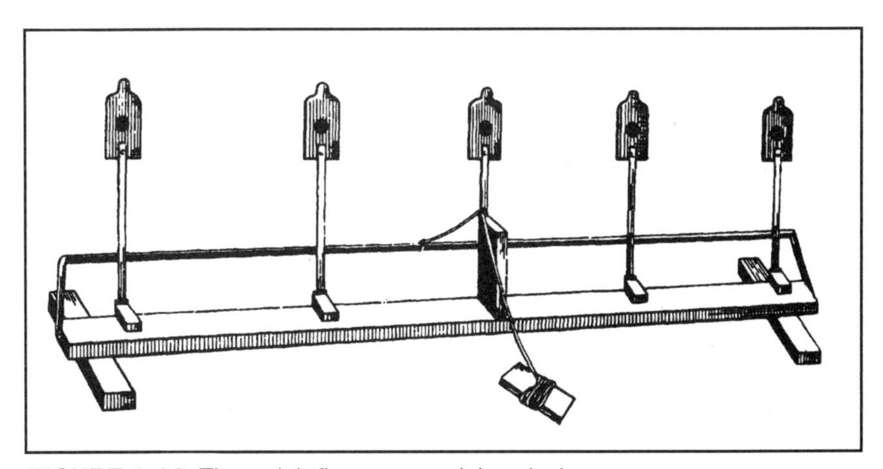

FIGURE 2-26. The quick-fire target training device.

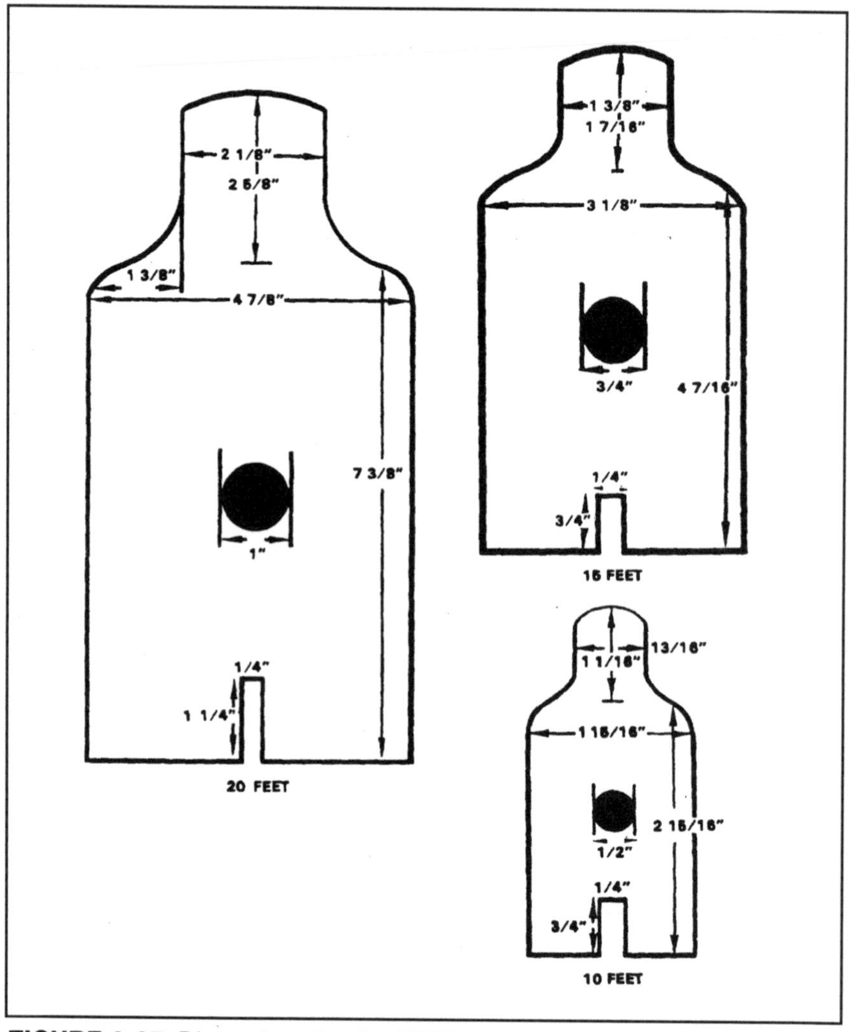

FIGURE 2-27. Dimensions for the QTTD.

PHASE IV. From 15 feet, six shots, at two 20-foot miniature E-type silhouettes.

(1) This exercise is conducted the same way as the previous one, except that the firer is introduced to *fire distribution.* The targets on the QTTD are held in the up position so they cannot be knocked down when hit.

(2) The firer first engages the 20-foot miniature E-type silhouette on the extreme right of the QTTD (see Figure 2-28). He then

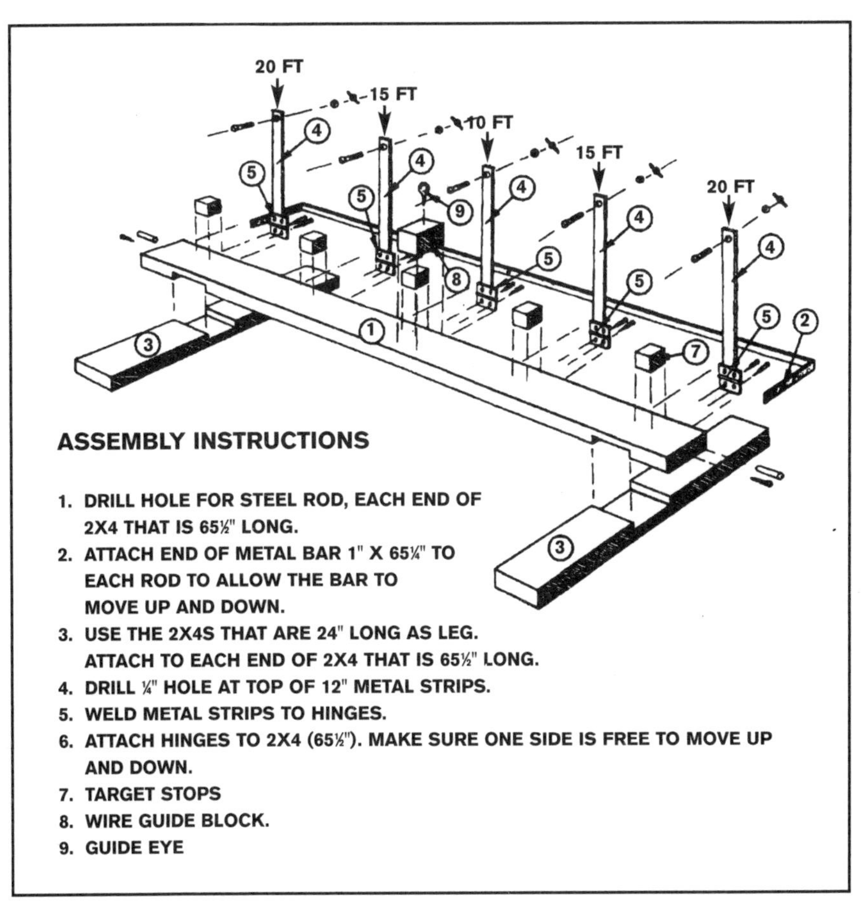

FIGURE 2-28. Miniature E-type silhouette for use with QTTD.

traverses between targets and engages the same type target on the extreme left of the QTTD. The firer again shifts back to reengage the first target. The procedure is used to teach the firer to instinctively return to the first target if he misses it with his first shot.

(3) The firer performs this exercise twice, firing three shots each time. Before firing the second time, the coach and firer should discuss the errors made during the first exercise.

PHASE V. Seven shots fired from 20, 15, and 10 feet at miniature E-type silhouettes.

(1) The firer starts this exercise 30 feet from the QTTD. The command, MOVE OUT, is given, and the firer steps out at a normal

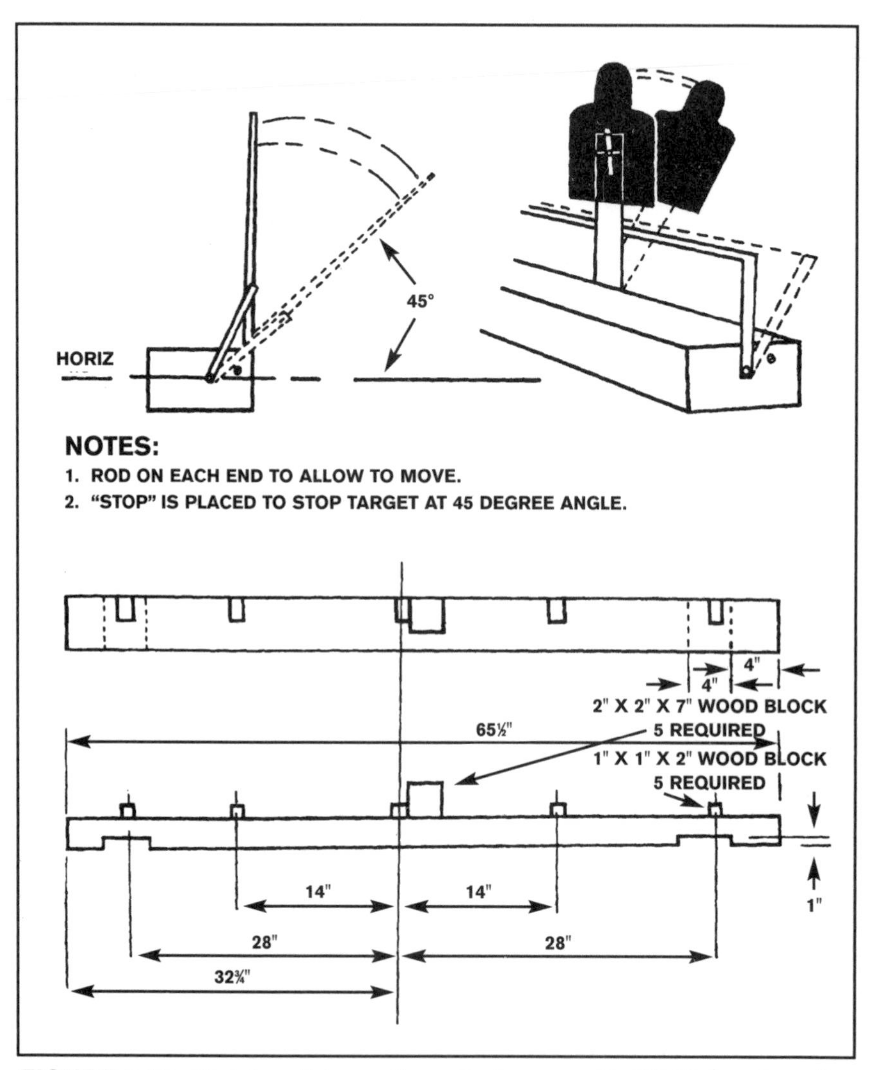

FIGURE 2-28. Miniature E-type silhouette for use with QTTD (continued).

pace with the weapon held in the ready position. Upon the command, FIRE (given at the 20-foot line), the firer assumes the crouch position and engages the 20-foot miniature E-type silhouette on the extreme right of the QTTD. He then traverses between targets, engages the same type target on the extreme left of the QTTD, and shifts back to the first target. If the target is still up, he engages it. The firer then assumes the standing position and returns the weapon to the ready position.

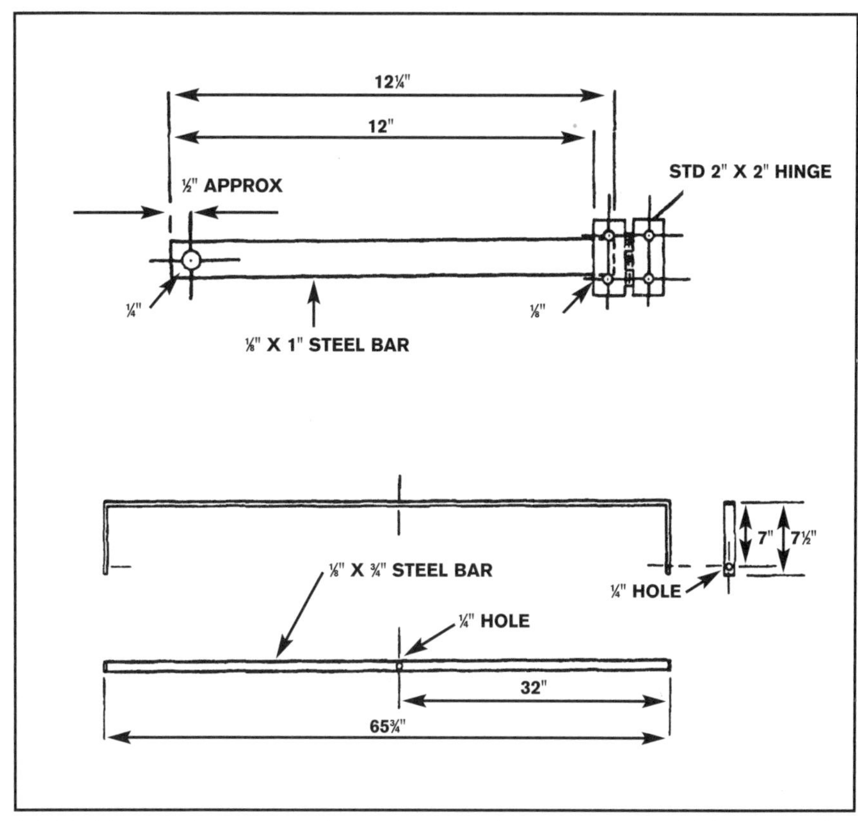

FIGURE 2-28. Miniature E-type silhouette for use with QTTD (continued).

Upon completion of each exercise, the coach makes corrections as the firer returns to the standing position.

(2) On the command, MOVE OUT, the firer again steps off at a normal pace. Upon the command, FIRE (given at the 15-foot line), he engages the 15-foot targets on the QTTD. The same sequence of fire distribution is followed as with the previous exercise.

(3) During this exercise, the firer moves forward on command, until he reaches the 10-foot line. At the command, FIRE, the firer engages the 10-foot miniature E-type silhouette in the center of the QTTD.

2-22. RANGE FIRING COURSES

Range firing is conducted after the firers have satisfactorily completed preparatory marksmanship training. The range firing courses are:

a. Instructional firing is practice firing on a range, using the assistance of a coach.

(1) All personnel authorized or required to fire the pistol or revolver receive 12 hours of preliminary instruction that includes the following:

- Disassembly and assembly (does not apply to revolver).
- Loading, firing, unloading, and immediate action.
- Preparatory marksmanship.
- Care and cleaning.

(2) The tables fired for instructional practice are prescribed in the combat pistol qualification course in Appendix A and in the revolver qualification course in Appendix C. During the instructional firing, the CPQC or RQC is fired with a coach or instructor.

NOTE: The RQC is fired on the same range as the CPQC; for a picture of the CPQC see FM 25-7.

b. The CPQC stresses the fundamentals of quick fire. It is the final test of a soldier's proficiency and the basis for his marksmanship classification. After the soldier has completed the instructional practice firing he will shoot the CPQC for record. A detailed description of the CPQC tables, standards, and conduct of fire is in Appendix A.

NOTE: The alternate pistol qualification course (APQC) or alternate revolver qualification course (ARQC) can be used for sustainment/qualification if the CPQC is not available (see Appendix B and Appendix D).

c. The military police firearms qualification course is a practical course of instruction for police firearms training (see FM 19-10).

SECTION IV.
SAFETY

Safety must be observed during all marksmanship training. Listed below are the precautions for each phase of training. These are not intended to replace AR 385-63 or local range regulations. Range safety requirements vary according to the requirements of the course of fire. It is mandatory that the latest range safety directives and local range regulations be consulted to determine current safety requirements.

2-23. REQUIREMENTS

a. A red flag is displayed prominently on the range during all firing.

b. Weapons must be handled carefully and are never pointed at anyone except the enemy in actual combat.

c. A weapon is always assumed loaded until it has been thoroughly examined and found to contain no ammunition.

d. Firing limits are indicated by red-and-white-striped poles visible to all firers.

e. Obstructions should never be placed in the muzzle of any weapon about to be fired.

f. Weapons are kept in a prescribed area with proper safeguards.

g. Smoking is not allowed on the range near ammunition, explosives, or flammables.

2-24. BEFORE FIRING

a. All prescribed roadblocks and barriers are closed, and guards are posted.

b. All weapons are checked to ensure they are clear of ammunition and obstructions, and slides are locked to the rear.

c. All firers are briefed on the firing limits of the range and firing lanes. They must keep their fires within prescribed limits.

d. All firers are instructed on how to load and unload the weapon, and on safety features.

e. All personnel are briefed on all safety aspects of fire and range pertaining to the conduct of the courses.

f. No one moves forward of the firing line without permission of the tower operator, safety officer, or OIC.

g. Weapons are loaded and unlocked only on command from the tower operator except during the conduct of the courses requiring automatic magazine changes.

h. Weapons are not handled except on command from the tower operator.

i. Firers must keep their weapons pointed downrange when loading, preparing to fire, or firing.

2-25. DURING FIRING

a. A firer does not move from his position until his weapon has been cleared by safety personnel, and it has been placed in its proper safety position. An exception is the assault phase.

b. During Table V of the CPQC, firers remain on line with other firers on their right or left.

c. Firers are careful to fire in their own firing lane and not to point the weapon into an adjacent lane, mainly during the assault phase.

d. The air-operated pistol is treated as a loaded weapon. Firers observe the same safety precautions as with other weapons.

e. All personnel wear helmets during live-fire exercises.

f. The weapon is held in the raised position except when preparing to fire. It is then held in the ready position, pointed downrange.

2-26. AFTER FIRING

a. Safety personnel inspect all weapons to ensure they are clear. A check is conducted to determine if any brass or live ammunition is in the possession of soldiers.

b. Once cleared, pistols are secured with the slides locked to the rear, and revolvers with cylinders open.

2-27. INSTRUCTIONAL PRACTICE AND RECORD QUALIFICATION FIRING

During these phases of firing, safety personnel ensure that the—

a. Firer understands the conduct of the exercise.

b. Firer has the required ammunition, and understands the commands for loading and unloading.

c. Firer complies with all commands from the tower operator.

d. Proper alignment is maintained with other firers while moving downrange.

e. Weapon is always pointed downrange.

f. Firer fires within the prescribed range limits.

g. Weapon is cleared after each phase of firing, and the tower operator is aware of the clearance.

h. Malfunction or failure to fire, due to no fault of the firer, is reported immediately. On command of the tower operator, the weapon is cleared and action is taken to allow the firer to continue with the exercise.

NOTE: For training and qualification standards see Appendixes A through E.

APPENDIX A

COMBAT PISTOL QUALIFICATION COURSE

A-1. COURSE INFORMATION

a. The CPQC may be used for both the pistol and revolver (for use with revolvers see Appendix C). It requires the soldier to engage single and multiple targets at various ranges using the fundamentals of quick fire. If the CPQC is not available, training may still be sustained and qualification achieved by using the APQC or ARQC.

NOTE: For a picture of the CPQC, see FM 25-7.

b. For each table of the CPQC, the firer is afforded extra rounds to reengage targets that are missed. During the course, 30 targets are exposed to the firer. However, 40 rounds of ammunition are issued with which to engage the exposed targets. A soldier who can reengage a target with an extra round during the exposure time is just as effective as one who hits the target with one round. The firer is not penalized for using or not using the extra ammunition. All excess ammunition is turned in after the completion of each table and may not be used during subsequent tables.

c. Two magazine changes are required when firing the CPQC. For safety purposes, one magazine contains one round of ammunition and is loaded first. A target appears in front of the firer, and he engages it. Eight seconds later, another target appears. During the eight-second delay, the firer must reload the weapon and be prepared to engage the next exposed target. There are no commands from range personnel or coaches for the magazine change. If the firer fails to reload his pistol in time to engage the next target, it is scored as a miss. This exercise teaches the soldier to quickly and safely change his magazine by instinct under pressure.

NOTE: When using the M9 pistol, the firer fires the first round in the double-action mode for all tables.

d. The range to exposed targets does not exceed 31 meters from the firer. Target exposure times are as follows:

(1) Tables I, II, and III:

(a) Single targets—three seconds.

(b) Multiple targets—five seconds.

(2) Tables IV and V:

(a) Single targets—two seconds.

(b) Multiple targets—four seconds.

A-2. FIRING THE CPQC

NOTE: The target sequence is decided by the tower operators but is the same for all *lanes. This prevents firers from getting ahead of firers in adjoining lanes. Target sequence will vary in distance from the firer, starting with 31 meters and allowing no more than two 7-meter targets.*

a. Qualification tables are as follows:

NOTE: Tower controls all reloading for revolvers.

(1) *Table I:* One magazine with seven rounds, and five targets exposed. The standing position is assumed at the firing line with the weapon held at the ready position. Only single targets are exposed to the firer in this table. Target sequence is decided by the tower operator.

(2) *Table II:* One magazine with one round, one magazine with seven rounds, and six targets exposed. The firer assumes the same position on the firing line as in Table I. There are four single targets and one set of multiple targets exposed to the firer.

(a) The magazine with one round is loaded into the weapon—one target is then exposed to the firer.

(b) After firing the pistol, the firer must change magazines at once. Three seconds after the target appears, the target is lowered if not hit.

(c) Eight seconds later, another target appears. Again, the firer must engage this target in the required time, or it is scored a miss.

(3) *Table III:* One magazine with seven rounds is loaded. Five targets are exposed—fired following rotation to another firing point. The firer assumes the same position on the firing line as in Tables I and II. Three single targets and one set of multiple targets are exposed to the firer. Target sequence is usually single, multiple, multiple, single, and multiple.

(4) *Table IV:* One magazine with five rounds is loaded. Four targets are exposed starting with the same position used in the previous tables. Two single targets and one set of multiple targets are exposed to the firer.

(5) *Table V:* One magazine with one round, one magazine with seven rounds, one magazine with five rounds, and ten targets exposed. The firer begins 10 meters behind the firing line in the middle of the trail.

(a) The magazine containing one round is loaded into the pistol. The firer places the magazine containing seven rounds in his magazine pouch where it is closest to the firing hand. The magazine containing five rounds is placed in the magazine pouch farthest from the firing hand.

(b) When the firer reaches the firing line, a single target is exposed for two seconds, then lowered if not hit. There is an eight-second delay to allow the firer to change magazines. The seven-round magazine should be loaded at this time.

(c) At the end of eight seconds, another single target is exposed to the firer. Again, should the firer not have loaded his second magazine in time to engage this target, it is scored a miss.

(d) When the tower operator is sure that the firing line has completed the magazine change, he gives the command, MOVE OUT.

He exposes two sets of multiple targets at various ranges from the firer.

(e) After exposure of two sets of multiple targets, the pistol is reloaded with the five-round magazine. The command, MOVE OUT, is given; and the remaining targets are presented to the firer in sequence. After the last targets are hit or lowered, the weapon is cleared.

(f) The firer, holding the weapon in the raised pistol position with the slide to the rear, returns to the starting point and places the weapon on the stand. Excess ammunition is turned in to the ammunition point. The next order moves to the firing line.

b. The same course is fired for night qualification. It is based on a GO/NO-GO scoring system: five target hits equal a GO. Ten seconds are allowed for each round.

c. The same course is fired for NBC qualification. It is based on a GO/NO-GO scoring system; seven target hits equal a GO. Ten seconds are allowed for each round.

NOTE: Night and NBC qualification is required IAW DA Pam 350-38.

A-3. CONDUCT OF FIRE

When the weapon is being fired, firers are issued the rounds required to fire a specific table. The following list of commands outlines a step-by-step sequence for conducting range firing on the CPQC.

a. *Table I.*

(1) The tower operator orders firers to move to the firing line in preparation for firing. The tower operator orders firers to position themselves next to the weapon stands and secure their weapons. Magazines containing seven rounds are issued to the scorers and given to the firers.

(2) The tower operator commands:

TABLE ONE, SEVEN ROUNDS.
LOAD AND LOCK.
READY ON THE RIGHT.
READY ON THE LEFT.
READY ON THE FIRING LINE.
UNLOCK YOUR WEAPONS.
WATCH YOUR LANE.

(3) The tower operator exposes the targets to the firers. When all targets have been exposed and engaged or lowered, the tower operator commands:

CEASE FIRE.
CLEAR ALL WEAPONS.
CLEAR ON THE RIGHT.
CLEAR ON THE LEFT.
THE FIRING LINE IS CLEAR.
FIRERS, PLACE YOUR WEAPONS ON THE STANDS.

b. *Table II.*

(1) The tower operator orders firers to secure their weapons. One magazine of one round and one magazine of seven rounds are issued to the firers.

(2) The tower operator commands:

TABLE TWO, EIGHT ROUNDS.
LOAD AND LOCK.
READY ON THE RIGHT.
READY ON THE LEFT.
READY ON THE FIRING LINE.
UNLOCK YOUR WEAPONS.
WATCH YOUR LANE.

(3) The tower operator exposes the targets to the firers. When all targets have been exposed and engaged or lowered, the tower operator commands:

CEASE FIRE.
CLEAR ALL WEAPONS.
CLEAR ON THE RIGHT.
CLEAR ON THE LEFT.
THE FIRING LINE IS CLEAR.
FIRERS, KEEP YOUR WEAPONS POINTED UP AND DOWNRANGE.
MOVE TO THE FIRING POINT TO YOUR RIGHT.

c. *Table III.*

(1) The tower operator orders the firers to position themselves next to the weapon stands. One magazine of seven rounds is issued to the firers.

(2) The tower operator commands:

TABLE THREE, SEVEN ROUNDS.
LOAD AND LOCK.
READY ON THE RIGHT.
READY ON THE LEFT.
READY ON THE FIRING LINE.
UNLOCK YOUR WEAPONS.
WATCH YOUR LANE.

(3) The tower operator exposes the targets to the firers. When all targets have been exposed and engaged or lowered, the tower operator commands:

CEASE FIRE.
CLEAR ALL WEAPONS.
CLEAR ON THE RIGHT.
CLEAR ON THE LEFT.
THE FIRING LINE IS CLEAR.
FIRERS, PLACE YOUR WEAPONS ON THE STAND.

d. *Table IV.*

(1) The tower operator orders the firers to secure their weapons and move to the center of the trail. Firers are issued one magazine of five rounds.

(2) The tower operator commands:

TABLE FOUR, FIVE ROUNDS.
LOAD AND LOCK.
READY ON THE RIGHT.
READY ON THE LEFT.
READY ON THE FIRING LINE.
UNLOCK YOUR WEAPONS.
WATCH YOUR LANE.

(3) The tower operator exposes the targets to the firers. When all targets have been exposed and engaged or lowered, the tower operator commands:

CEASE FIRE.
CLEAR ALL WEAPONS.
CLEAR ON THE RIGHT.
CLEAR ON THE LEFT.

THE FIRING LINE IS CLEAR.

FIRERS, PLACE YOUR WEAPONS ON THE STAND TO THE REAR OF THE FIRING LINE.

e. *Table V.*

(1) The tower operator orders the firers to secure their weapons. Firers are issued one magazine of one round, one magazine of seven rounds, and one magazine of five rounds.

(2) The tower operator commands:

TABLE FIVE, THIRTEEN ROUNDS.

LOAD AND LOCK.

READY ON THE RIGHT.

READY ON THE LEFT.

READY ON THE FIRING LINE.

PISTOLS AT THE READY POSITION.

UNLOCK YOUR WEAPONS.

WATCH YOUR LANE.

MOVE OUT.

(3) The tower operator exposes the targets to the firers. He gives the firers the commands, WEAPONS AT THE READY POSITION and MOVE OUT, after each target or group of targets has been engaged.

(4) Upon completion of Table V, the tower operator commands:

CEASE FIRE.

CLEAR ALL WEAPONS.

CLEAR ON THE RIGHT.

CLEAR ON THE LEFT.

THE FIRING LINE IS CLEAR.

FIRERS, KEEP YOUR WEAPONS UP AND DOWNRANGE.

SCORERS AND FIRERS, MOVE BACK TO THE FIRING LINE AND PLACE YOUR WEAPONS ON THE STAND.

(5) The tower operator has each scorer total the firer's scorecard and turn it in to the range officer or his representative. The firing orders are rotated and the above sequence continued until all orders have fired.

NOTE: For night qualification and NBC qualification, the same course is used. Ten seconds is allowed for each round.

A-4. ALIBIS

a. Alibis are fired at the completion of each table from the position where the alibi occurred. Fire commands that apply to the table are used to fire an alibi.

b. If a malfunction of the weapon or target occurs during firing from stationary positions, the firer reports the malfunction, and keeps his weapon pointed up and downrange. Should the malfunction occur during Table V, the firer keeps his weapon pointed up and downrange. He continues to move forward, keeping aligned with the firers to his right and left.

A-5. RULES

Rules governing firing the CPQC are as follows:

a. Coaching. Coaching is not allowed during record firing. No person may give or try to give help while the firer is taking his position or after he has taken his position at the firing point. Each firer must observe the location of the target in his own lane. During the instructional firing, the coach and assistant instructors should assist the firer in correcting errors.

b. Accidental Discharges. All shots fired by the firer are scored after he has taken his place on the firing lane. Even if the weapon is not directed toward a target or is accidentally discharged, a replacement round is not issued.

c. Firing on the Wrong Target. Shots fired on the wrong target are entered as a miss on the firing scorecard. A firer is credited with hits he attains in his own firing lane.

d. Firing After the Signal to Lower Targets. Any shot fired by a firer after targets start to lower is scored as a miss.

e. More Than One Shot Fired at an E-Type Silhouette Target. The firer is credited with a hit if the hit is made during the target exposure time. The number of rounds fired to obtain the hit is immaterial.

f. Excess Ammunition at the End of a Firing Table. Excess ammunition from each table is turned in to the ammunition point and not used by the firer for subsequent tables.

g. Target Sequence. The target sequence is decided by the tower operator but is the same for all lanes. This prevents firers from getting ahead of firers in adjoining lanes. Target sequence will vary in distance from the firer, starting with 31 meters and allowing no more than two 7-meter targets.

A-6. SCORECARD

a. Use. The scorecard outlines instructional firing and qualification firing (CPQC) (see Figure A-1). Numbers in columns labeled TGT (target) are not the sequence in which targets are exposed. They are the numerical identification of targets to be engaged during each table of fire.

b. Scoring. Each time a target is hit or "killed", an X is placed in the column labeled HITS. The value of a hit is 10 points. Upon completion of firing the CPQC, the scorer totals and signs the scorecard. Qualification standards are listed in the bottom right corner on the record firing side of the scorecard. They are:

Expert	260 to 300
Sharpshooter	210 to 250
Marksman	160 to 200
Unqualified	below 160

NBC and night firing are done on a GO/NO-GO scoring system and recorded in the remarks column.

NBC: 7 target hits = GO

Night: 5 target hits = GO

c. Supply of Forms. DA Form 88 is available through normal publications supply channels.

A-7. TARGETS

Seven electric target device targets and E-type silhouettes for each firing lane are required. Aggressor figures may be superimposed on the silhouettes to add realism to the course of fire.

COMBAT PISTOL QUALIFICATION COURSE SCORECARD

FOR USE OF THIS FORM SEE FM 23-35. PROPONENT AGENCY IS USCONARC.

Lane No. 1 | Order 2 | Group 2 | Date 28 MAR 88 | Unit C Co 2/29

Name Fetty (Last) Leonard (First) H. (MI) 232-00-6380 (SSN)

TABLE I

1 magazine 7 rds.

TIME	TGT	HITS
3 Seconds	1	X
3 Seconds	2	X
3 Seconds	3	M
3 Seconds	4	X
3 Seconds	5	X
TOTAL		40

TABLE II

1 magazine 1 rd.
8 second delay for magazine change.
1 magazine 7 rds.

TIME	TGT	HITS
3 Seconds	1	X
3 Seconds	2	X
5 Seconds	3	X
	4	X
3 Seconds	5	X
3 Seconds	6	X
TOTAL		60

TABLE III

1 magazine 7 rds.

TIME	TGT	HITS
3 Seconds	1	X
3 Seconds	2	X
3 Seconds	3	X
5 Seconds	4	X
	5	X
TOTAL		50

TABLE IV

1 magazine 5 rds.

TIME	TGT	HITS
2 Seconds	1	M
2 Seconds	2	X
4 Seconds	3	X
	4	X
TOTAL		30

TABLE V

1 magazine 1 rd.
8 second delay for magazine change.
1 magazine 7 rds.
Controlled magazine change 1 magazine 8 rds.

TIME	TGT	HITS
2 Seconds	1	X
2 Seconds	2	X
4 Seconds	3	M
	4	X
4 Seconds	5	X
	6	X
2 Seconds	7	X
2 Seconds	8	X
4 Seconds	9	X
	10	X
TOTAL		90

TOTAL HITS	27
TOTAL SCORE	270

Scorer's Signature

Officer's Signature

QUALIFICATION

Expert	260
Sharpshooter	210
Marksman	160

Notes: 1. When the revolver is being fired, firers are issued number of rounds required to fire a specific table. Officer in charge of firing establishes procedure for loading and reloading. All reloading is controlled; however, time allowed for target engagement is not changed.
2. Tables 1, 2, 3 are fired for familiarization and instructional firing.

DA FORM 88, Sep 71 — Previous editions of this form are obsolete.

FIGURE A-1. Sample scorecard, DA Form 88.

A-8. QUICK-FIRE TARGET TRAINING DEVICE

The QTTD may be procured locally. For durability and appearance, it should be made by the training aids section or an equally capable agency.

APPENDIX B

ALTERNATE PISTOL QUALIFICATION COURSE

B-1. PROCEDURES

Once the soldier has completed instructional firing, he must then fire the CPQC for record. If the CPQC is not available, the soldier can fire the APQC.

a. Procedures for firing the APQC are as follows. Given 40 rounds of ammunition, fire Tables 1 through 4.

(1) *Table 1:* Engage the 25-meter APQC target from the standing position with 7 rounds of ammunition; given one 7-round magazine on a 25-meter range during daylight hours. Within 21 seconds engage the APQC target from the standing position.

(2) *Table 2:* Engage the 25-meter APQC target from the kneeling position with 13 rounds; given two magazines, one 6-round and one 7-round, on a 25-meter range during hours of daylight. Within 45 seconds, from a standing position, assume a good kneeling position, engage the target with 6 rounds, perform a rapid magazine change, and engage the target with a 7-round magazine.

(3) *Table 3:* Engage the 25-meter APQC target from the crouch position with 10 rounds; given two magazines with 5 rounds each on a 25-meter range during daylight hours. Within 35 seconds, from a standing position, assume a good crouch position, engage the target with one 5-round magazine, perform a rapid magazine change, and engage the target with the second 5-round magazine.

(4) *Table 4:* Engage the 25-meter APQC target from the prone

position with 10 rounds; given two magazines with 5 rounds each on a 25-meter range during daylight hours. Within 35 seconds, from a standing position, assume a good prone position, engage the target with one 5-round magazine, perform a rapid magazine change, and engage the target with the second 5-round magazine.

b. Firing Pistol Under Night Conditions. Engage the 25-meter target from the crouch position with 30 rounds; given two 15-round magazines of M9 9-mm ammunition or four 7-round magazines and one 2-round magazine of M1911A1 ammunition on a 25-meter range under night conditions. Given 10 seconds for each round, engage E-type silhouettes with 10 rounds. Conduct magazine changes without command. Tower will allow 8 seconds for each magazine change.

c. Firing Pistol Under NBC Conditions. Engage a 25-meter target from a crouch position with 20 rounds; given one 15-round magazine and one 5-round magazine of M9 9-mm ammunition or two 7-round magazines and one 6-round magazine of M1911A1 ammunition on a 25-meter range under simulated NBC conditions. During daylight hours, given 10 seconds for each round, engage E-type silhouettes with 20 rounds of ammunition. Conduct magazine changes without command. Tower will allow 10 seconds for each magazine change.

NOTE: When using the 9-mm pistol, the first round is fired in the double-action mode for all four tables. Night and NBC qualification is required IAW DA Pam 350-38.

B-2. CONDUCT OF FIRE

a. The following commands outline a step-by-step sequence for conducting range firing on the APQC.

(1) *Table 1:* Standing position.

(a) The tower operator gives the order to move to the firing line and to prepare to fire. The magazine containing seven rounds is issued to the scorer and given to the firer on command. The tower operator commands:

TABLE ONE, STANDING POSITION, SEVEN ROUNDS.
LOAD AND LOCK.
IS THE LINE READY?

(The 9-mm firers place their weapons in the double-action mode at this time.)

THE FIRING LINE IS READY.

FIRERS, WATCH YOUR LANE!

(b) At the end of the prescribed firing time, the tower operator commands:

CEASE FIRE.

ARE THERE ANY ALIBIS?

(Alibis are given eight seconds for each round not fired.)

NOTE: For more information see paragraph B-3.

UNLOAD AND CLEAR ALL WEAPONS.

IS THE FIRING LINE CLEAR?

THE FIRING LINE IS NOW CLEAR.

FIRERS AND SCORERS, MOVE DOWNRANGE AND CHECK YOUR TARGETS.

(Weapons are left on the firing line with slides locked to the rear.)

NOTE: Clear, lock open, and leave weapons on the table, or stand weapons at the firing line when the firer and scorer go downrange to score their target.

(2) *Table 2:* Kneeling position.

The tower operator orders firers to move up to the firing line. Two magazines containing six rounds and seven rounds each are issued to the scorer to be given to the firer on command. The tower operator commands:

TABLE TWO, KNEELING POSITION WITH MAGAZINE CHANGE, FORTY-FIVE SECONDS.

LOCK AND LOAD ONE SIX-ROUND MAGAZINE.

LOAD YOUR SEVEN-ROUND MAGAZINE WITHOUT COMMAND.

NOTE: The following commands are the same as for Table 1.

(3) *Table 3:* Crouch position.
The tower operator orders firers to move up to the firing line. Scorers are issued two 5-round magazines to be issued to the firer on command. The tower operator commands:

TABLE THREE, CROUCH POSITION WITH MAGAZINE CHANGE, THIRTY-FIVE SECONDS.

LOAD YOUR SECOND FIVE-ROUND MAGAZINE WITHOUT COMMAND.

NOTE: The following commands are the same as for Tables 1 and 2.

(4) *Table 4:* Prone position.
The tower operator orders firers to move to the firing line. Firers are issued two 5-round magazines. The tower operator orders:

TABLE FOUR, PRONE POSITION WITH MAGAZINE CHANGE, THIRTY-FIVE SECONDS.

LOAD YOUR SECOND FIVE-ROUND MAGAZINE WITHOUT COMMAND.

NOTE: The following commands are the same as for Tables 1, 2, and 3.

(5) The scorer and firer repair or replace targets for the next firing order.

b. The commands for the pistol night fire for record are as follows:

(1) The tower operator orders to move to the firing line and to prepare to fire. Two magazines of 15 rounds of M9 ammunition or four 7-round magazines and one 2-round magazine of M1911A1 ammunition are issued to firers.

(2) The tower operator commands:

NIGHT FIRE, CROUCH POSITION WITH MAGAZINE CHANGES.

LOAD OTHER MAGAZINES WITHOUT COMMAND.

LOAD AND LOCK ONE MAGAZINE.

(M1911A1 firers must load their two-round magazine first.)

IS THE FIRING LINE READY?

(M9 firers must place their weapons in the double-action mode.)

THE FIRING LINE IS READY.

FIRERS, WATCH YOUR LANE.

(3) At the end of the prescribed firing time, the tower operator commands:

CEASE FIRE.

ARE THERE ANY ALIBIS?

(Alibis are given 10 seconds for each round not fired.)

UNLOAD AND CLEAR ALL WEAPONS.

IS THE FIRING LINE CLEAR?

THE FIRING LINE IS NOW CLEAR.

FIRERS AND SCORERS, MOVE DOWNRANGE AND CHECK YOUR TARGETS.

(Weapons are left on the firing line with slides locked to the rear.)

c. The commands for the pistol NBC fire for record are as follows:

(1) The tower operator orders to move to the firing line and to prepare to fire. Firer is given one 15-round magazine and one 5-round magazine of M9 ammunition or two 7-round magazines and one 6-round magazine of M1911A1 ammunition.

(2) The tower operator commands:

NBC FIRE, CROUCH POSITION WITH MAGAZINE CHANGE.

LOAD OTHER MAGAZINES WITHOUT COMMAND.

LOAD AND LOCK ONE MAGAZINE.

(M9 firers load 5-round magazine first; M1911A1 firers load 6-round magazine first.)

IS THE FIRING LINE READY?

(M9 firers must place their weapons in the double-action mode.)

THE FIRING LINE IS READY.

FIRERS, WATCH YOUR LANE.

(3) At the end of the prescribed firing time, the tower operator commands:

CEASE FIRE.

ARE THERE ANY ALIBIS?

(Alibis are given eight seconds for each round not fired.)

UNLOAD AND CLEAR ALL WEAPONS.

IS THE FIRING LINE CLEAR?

THE FIRING LINE IS NOW CLEAR.

FIRERS AND SCORERS, MOVE DOWNRANGE AND CHECK YOUR TARGETS.

(Weapons are left on the firing line with slides locked to the rear.)

NOTE: Excess ammunition at the end of a firing table is turned in to the scorer and not used by the firer in subsequent tables. At the end of the course, all excess ammunition is turned in to the ammunition point.

B-3. ALIBIS

If there is a malfunction of the weapon or target during firing, the scorer reports and records the malfunction. The firer is allowed one alibi (eight seconds for each round) at the completion of each table. All alibis are fired from the position in which the alibi occurred. Firing commands that apply are used to fire alibis.

B-4. SCORING

a. The firer is scored on the number of target hits during the time limit. The firer must achieve at least 24 hits with a minimum score of 80 points to qualify. The target hits are multiplied by the number inside the scoring rings to determine the score. No credit is given for rounds fired after the command CEASE FIRE. Shots that touch the next higher scoring ring are scored the next higher value. (See Figure B-1.)

b. The qualification scores are:

Expert	160 to 200
Sharpshooter	120 to 159
Marksman	80 to 119

NBC and night firing are done on a GO/NO-GO scoring system and recorded in the remarks column.

NBC: 7 target hits = GO

Night: 5 target hits = GO

NOTE: See format for scorecard in Figure B-2.

c. Coaching is allowed during instructional firing but not during record fire. No one may assist while the firer is taking position or after taking position at the firing point except for safety reasons.

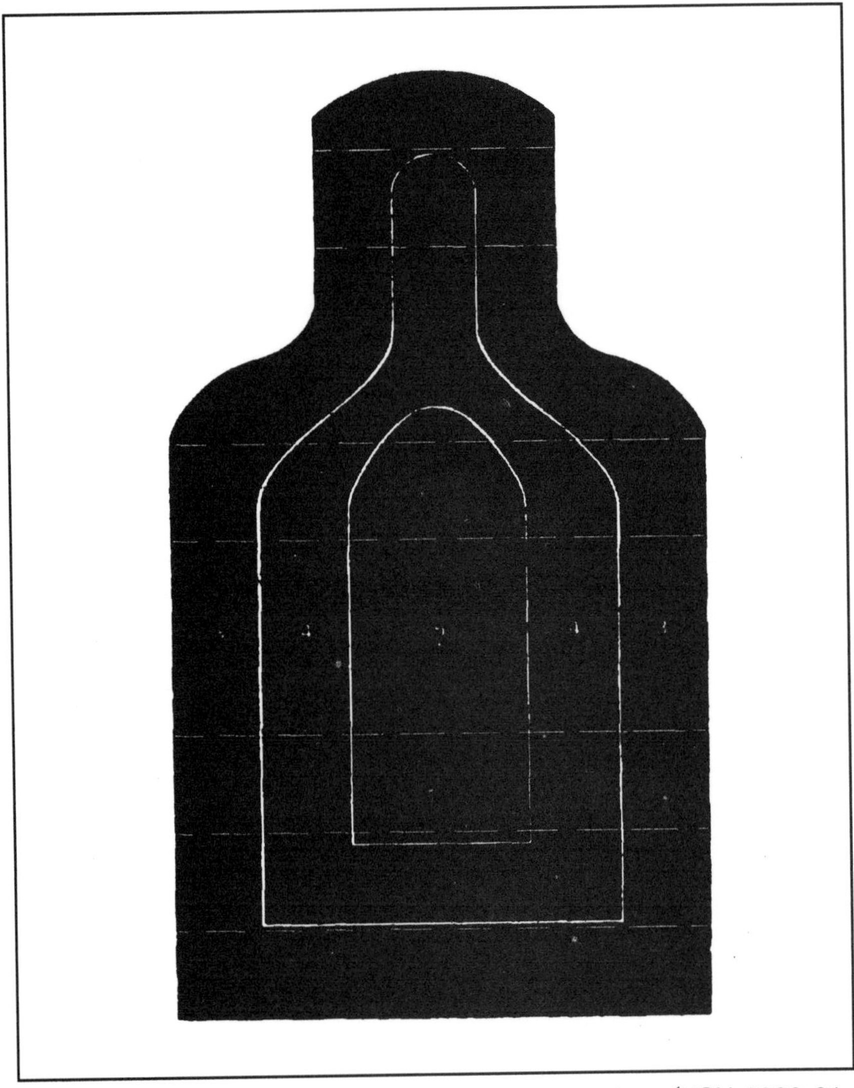

FIGURE B-1. The 25-meter E-type silhouette with rings (NSN 6920-01-276-6604).

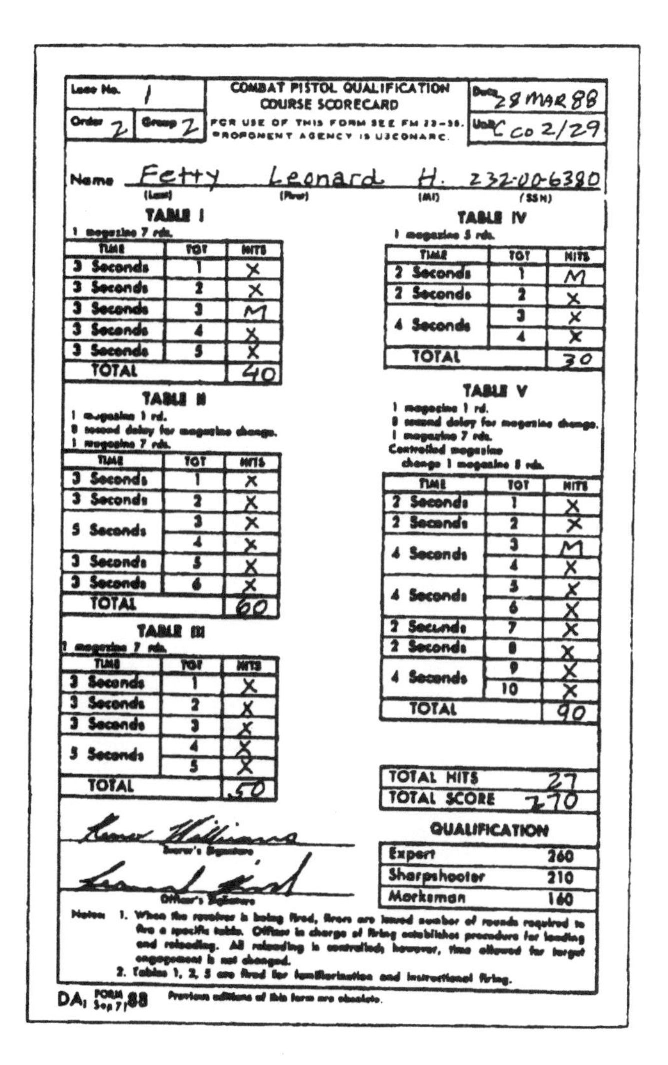

COMBAT PISTOL QUALIFICATION COURSE SCORECARD

FOR USE OF THIS FORM SEE FM 23-35. PROPONENT AGENCY IS USCONARC.

Lane No. 1 | Order 2 | Group 2 | Date 28 MAR 88 | Unit C CO 2/29

Name Fetty (Last) Leonard (First) H. (MI) 232-00-6380 (SSN)

TABLE I

1 magazine 7 rds.

TIME	TGT	HITS
3 Seconds	1	X
3 Seconds	2	X
3 Seconds	3	M
3 Seconds	4	X
3 Seconds	5	X
TOTAL		40

TABLE II

1 magazine 1 rd.
8 second delay for magazine change.
1 magazine 7 rds.

TIME	TGT	HITS
3 Seconds	1	X
3 Seconds	2	X
5 Seconds	3	X
	4	X
3 Seconds	5	X
3 Seconds	6	X
TOTAL		60

TABLE III

1 magazine 7 rds.

TIME	TGT	HITS
3 Seconds	1	X
3 Seconds	2	X
3 Seconds	3	X
5 Seconds	4	X
	5	X
TOTAL		50

TABLE IV

1 magazine 5 rds.

TIME	TGT	HITS
2 Seconds	1	M
2 Seconds	2	X
4 Seconds	3	X
	4	X
TOTAL		30

TABLE V

1 magazine 1 rd.
8 second delay for magazine change.
1 magazine 7 rds.
Controlled magazine change 1 magazine 5 rds.

TIME	TGT	HITS
2 Seconds	1	X
2 Seconds	2	X
4 Seconds	3	M
	4	X
4 Seconds	5	X
	6	X
2 Seconds	7	X
2 Seconds	8	X
4 Seconds	9	X
	10	X
TOTAL		90

TOTAL HITS	27
TOTAL SCORE	270

QUALIFICATION

Expert	260
Sharpshooter	210
Marksman	160

Scorer's Signature

Officer's Signature

Notes: 1. When the revolver is being fired, firers are issued number of rounds required to fire a specific table. Officer in charge of firing establishes procedure for loading and reloading. All reloading is controlled; however, time allowed for target engagement is not changed.
2. Tables 1, 2, 5 are fired for familiarization and instructional firing.

DA FORM 88, Sep 71 Previous editions of this form are obsolete.

FIGURE B-2. Example of completed Alternate Pistol Qualification Course form.

NOTE: See Appendix F for blank copy of this form for local reproduction.

APPENDIX C

REVOLVER QUALIFICATION COURSE

C-1. COURSE INFORMATION

a. The CPQC is used for both pistols and revolvers. This appendix outlines its use with revolvers only (for use with pistols see Appendix A). The CPQC requires the soldier to engage single and multiple timed targets at various ranges using the fundamentals of quick fire. If a CPQC is not available, training and qualification may be conducted using the standard 25-meter range and the ARQC (see Appendix D).

NOTE: For range design and layout of the CPQC, see FM 25-7.

b. For each table of the RQC, the firer is afforded extra rounds to reengage targets that are missed. During the course 30 targets are presented to the firer; however, the firer is given 40 rounds to engage these targets. A firer who can successfully reengage the target with a second round during the exposure time is just as effective as a firer who hits the target with the first round. The firer is not penalized for using or not using the extra rounds he is allocated. All excess ammunition is turned in at the end of each table and is not used for subsequent tables.

c. All reloads will be controlled by the tower operator. If the firer fails to engage a target within the timed exposure, that target is scored as a miss. This teaches him to quickly apply the fundamentals of pistol marksmanship under stress.

d. The range to exposed targets does not exceed 31 meters from the firer. Target exposure times are as follows:

(1) Tables I, II, and III:
(a) Single targets—three seconds.
(b) Multiple targets—five seconds.
(2) Tables IV and V:
(a) Single targets—two seconds.
(b) Multiple targets—four seconds.

C-2. FIRING THE RQC

NOTE: Target sequence is decided by the tower operator, but is the same for all lanes to prevent firers from getting in front of other firers in adjoining lanes. Targets will vary in distance to the firers, starting at 31 meters and allowing no more than two 7-meter targets.

a. Qualification tables are as follows:

(1) *Table I:* The revolver is loaded with six rounds. The standing position is assumed at the firing line with the weapon in the ready position. Four targets are exposed. The tower operator controls the reloading of the last round, followed by the exposure of the last target. Firers are reminded before the beginning of the table that they will have only seven rounds for five targets.

(2) *Table II:* The revolver is loaded with six rounds. Two single and one set of multiple targets are exposed before reloading is conducted under control of the tower operator. The remaining two rounds are loaded, and the last two single targets are exposed. The firer is advised before the start of the table that he will have only eight rounds with which to engage the six targets. Firers assume the same position as in Table I.

(3) *Table III:* The revolver is loaded with six rounds. One single and one set of multiple targets are exposed, followed by the reloading of the last round under the control of the tower operator. The remaining two single targets are then exposed to the firer. Firers are reminded before the start of the table that they will have seven rounds to engage five targets.

(4) *Table IV:* The revolver is loaded with five rounds. Two

single and one multiple target are exposed to the firer. No reloading takes place in this table.

(5) *Table V:* Firers are given 13 rounds. Ten targets are exposed throughout the table. The firer begins 10 meters behind the firing line in the middle of the trail.

(a) Six rounds are loaded into the revolver.

(b) When the firer reaches the firing line, a single target is exposed for two seconds, then lowered if not hit.

(c) One set of multiple targets is exposed to the firer. The firer is allowed four seconds to engage the targets. If targets are not engaged, they are scored a miss.

(d) When the tower operator has controlled reloading, he gives the command, MOVE OUT, and exposes two sets of multiple targets at various ranges from the firer.

(e) When the tower operator has controlled reloading, he gives the command MOVE OUT, and the remaining targets are presented in sequence. After the last targets are hit or lowered, the weapon is cleared.

(f) The firer, holding the weapon in the raised position with the cylinders open, returns to the starting point and places the weapon on the stand. Excess ammunition (if any) is turned in to the ammunition point. The next order moves to the firing line.

b. The same course is fired for night qualification. It is scored on a GO/NO-GO scoring system: five target hits equal a GO. Ten seconds are allowed for each round.

c. The same course is fired for NBC qualification. It is based on a GO/NO-GO system: seven target hits equal a GO. Ten seconds are allowed for each round.

NOTE: Night and NBC qualification is required IAW DA Pam 350-38.

C-3. CONDUCT OF FIRE

When the weapon is being fired, firers are issued the number of rounds required to fire a specific table. The tower operator controls all loading and reloading. The following list of commands outlines a step-by-step sequence for conducting range firing on the RQC.

a. *Table I.*

(1) The tower operator orders firers to move to the firing line in preparation for firing. The tower operator orders firers to position themselves next to the weapon stands and secure their weapons. Seven rounds are issued to scorers to be given to firers.

(2) The tower operator commands:

TABLE ONE, SEVEN ROUNDS.
LOAD SIX ROUNDS.
READY ON THE RIGHT.
READY ON THE LEFT.
READY ON THE FIRING LINE.
WATCH YOUR LANE.

(3) The tower operator exposes two single targets to the firers. Once these targets have been engaged or lowered, the tower operator commands:

CEASE FIRE.
LOAD REMAINING ROUND. (Tower allows appropriate time.)
READY ON THE RIGHT.
READY ON THE LEFT.
READY ON THE FIRING LINE.
WATCH YOUR LANE.

(4) The tower operator exposes the remaining three single targets to the firers. When all targets have been engaged or lowered, the tower operator commands:

CEASE FIRE.
CLEAR ALL WEAPONS.
CLEAR ON THE RIGHT.
CLEAR ON THE LEFT.
THE FIRING LINE IS CLEAR.
FIRERS, PLACE YOUR WEAPONS ON THE STANDS.

(Leave cylinders open.)

b. *Table II.*

(1) The tower operator orders firers to secure their weapons.

Eight rounds are issued to the scorers to be given to the firers.

(2) The tower operator commands:

TABLE TWO, EIGHT ROUNDS.

LOAD SIX ROUNDS.

READY ON THE RIGHT.

READY ON THE LEFT.

READY ON THE FIRING LINE.

WATCH YOUR LANE.

(3) The tower operator exposes four single targets to the firers. When these targets have been engaged or lowered, the tower operator commands:

CEASE FIRE.

LOAD TWO REMAINING ROUNDS. (Tower allows appropriate time.)

READY ON THE RIGHT.

READY ON THE LEFT.

READY ON THE FIRING LINE.

WATCH YOUR LANE.

(4) The tower operator exposes one set of multiple targets to firers. Once these targets have been engaged or lowered, the tower operator commands:

CEASE FIRE.

CLEAR ALL WEAPONS.

CLEAR ON THE RIGHT.

CLEAR ON THE LEFT.

THE FIRING LINE IS CLEAR.

FIRERS, KEEP YOUR WEAPONS POINTED UP AND DOWNRANGE.

MOVE TO THE FIRING POINT TO YOUR RIGHT.

c. *Table III.*

(1) The tower operator orders the firers to position themselves next to the weapon stands and secure their weapons. Seven rounds are issued to the scorers to be given to the firers.

(2) The tower operator commands:

TABLE THREE, SEVEN ROUNDS.

READY ON THE RIGHT.

READY ON THE LEFT.

READY ON THE FIRING LINE.

WATCH YOUR LANE.

(3) The tower operator exposes three single targets to the firers. When all targets have been engaged or lowered, the tower operator commands:

CEASE FIRE.

LOAD REMAINING ROUND. (Tower allows appropriate time.)

READY ON THE RIGHT.

READY ON THE LEFT.

READY ON THE FIRING LINE.

WATCH YOUR LANE.

(4) The tower operator exposes one set of multiple targets to the firers. When all targets have been engaged or lowered, the tower operator commands:

CEASE FIRE.

CLEAR ALL WEAPONS.

CLEAR ON THE RIGHT.

CLEAR ON THE LEFT.

THE FIRING LINE IS CLEAR.

FIRERS, PLACE YOUR WEAPONS ON THE STAND.

(Leave cylinders open.)

d. *Table IV.*

(1) The tower operator orders the firers to secure their weapons from the weapon stand and move to the center of the trail. Scorers are issued five rounds to be given to the firers.

(2) The tower operator commands:

TABLE FOUR, FIVE ROUNDS.

LOAD FIVE ROUNDS.

READY ON THE RIGHT.

READY ON THE LEFT.

READY ON THE FIRING LINE.

WATCH YOUR LANE.

(3) The tower operator exposes two single targets and one set of multiple targets to the firers. When all targets have been engaged or lowered, the tower operator commands:

CEASE FIRE.

CLEAR ALL WEAPONS.

CLEAR ON THE RIGHT.

CLEAR ON THE LEFT.

THE FIRING LINE IS CLEAR.

FIRERS, PLACE YOUR WEAPONS ON THE STANDS TO THE REAR OF THE FIRING LINE.

(Leave cylinders open.)

e. *Table V.*

(1) The tower operator orders the firers to secure their weapons. Seven are given 13 rounds to be given to the firers.

(2) The tower operator commands:

TABLE FIVE, THIRTEEN ROUNDS.

LOAD SIX ROUNDS.

READY ON THE RIGHT.

READY ON THE LEFT.

READY ON THE FIRING LINE.

WEAPONS AT THE READY POSITION.

WATCH YOUR LANE.

MOVE OUT.

(3) The tower operator exposes one single target, then one set of multiple targets to the firers. Once the targets have been engaged or lowered, the tower operator commands:

CEASE FIRE.

RELOAD CHAMBERS. (Tower operator allows appropriate time.)

READY ON THE RIGHT.

READY ON THE LEFT.

READY ON THE FIRING LINE.

WEAPONS IN THE READY POSITION.

WATCH YOUR LANE.

MOVE OUT.

(4) The tower operator exposes two sets of multiple targets to the firers. Once the targets have been engaged or lowered, the tower operator commands:

CEASE FIRE.
RELOAD CHAMBERS.
READY ON THE RIGHT.
READY ON THE LEFT.
READY ON THE FIRING LINE.
WEAPONS IN THE READY POSITION.
WATCH YOUR LANE.
MOVE OUT.

(5) The tower operator exposes one set of multiple targets and one single target to the firers. Once the targets have been engaged or lowered, the tower operator commands:

CEASE FIRE.
CLEAR ALL WEAPONS.
CLEAR ON THE RIGHT.
CLEAR ON THE LEFT.
THE FIRING LINE IS CLEAR.
FIRERS, KEEP YOUR WEAPONS UP AND DOWN-RANGE.
SCORERS AND FIRERS, MOVE BACK TO THE FIRING LINE AND PLACE YOUR WEAPONS ON THE WEAPON STANDS.

(Leave cylinders open.)

(6) The tower operator has each scorer total the firer's score-card and turn it in to the range officer or his representative. The firing orders are rotated and the above sequence continued until all orders have fired.

NOTE: For night qualification and NBC qualification, the same course is used. Ten seconds is allowed for each round.

C-4. ALIBIS

a. Alibis are fired at the completion of each table from the position where the alibi occurred. Fire commands that apply to the table are used to fire the alibi.

b. If a malfunction of the weapon or target occurs during firing from stationary positions, the firer reports the malfunction and keeps his weapon pointed up and downrange. Should the malfunction occur during Table V, the firer keeps his weapon pointed up and downrange. He continues to move forward, keeping aligned with the firers to his right and left.

C-5. RULES

Rules governing firing the CPQC are as follows:

a. Coaching. Coaching is not allowed during record firing. No person may give or try to give help while the firer is taking his position or after he has taken his position at the firing point. Each firer must observe the location of the target in his own lane. During the instructional firing, the coach and assistant instructors should assist the firer in correcting errors.

b. Accidental Discharges. All shots fired by the firer are scored after he has taken his place on the firing lane. Even if the weapon is not directed toward a target or is accidentally discharged, a replacement round is not issued.

c. Firing on the Wrong Target. Shots fired on the wrong target are entered as a miss on the firing scorecard. A firer is credited with hits he attains in his own firing lane.

d. Firing After the Signal to Lower Targets. Any shot fired by a firer after targets start to lower is scored as a miss.

e. More Than One Shot Fired at an E-Type Silhouette Target. The firer is credited with a hit if the hit is made during the target exposure time. The number of rounds fired to obtain the hit is immaterial.

f. Excess Ammunition at the End of a Firing Table. Excess ammunition from each table is turned in to the ammunition point and not used by the firer for subsequent tables.

g. Rounds Issued. Firers are issued the number of rounds required to fire a specific table.

h. Target Sequence. Target sequence is controlled by the tower operator but is the same for *all* lanes to prevent firers from getting in front of firers in adjoining lanes. Targets vary in distance from the firer, starting with 31 meters and allowing no more than two 7-meter targets.

C-6. SCORECARD

a. Use. The scorecard (DA Form 88) outlines instructional firing and qualification firing (CPQC) (see Figure A-1). Numbers in columns labeled TGT (target) are not the sequence in which targets are exposed. They are the numerical identification of targets to be engaged during each firing table.

NOTE: DA Form 88 is used to score the revolver qualification course.

b. Scoring. Each time a target is hit or "killed," an X is placed in the column labeled HITS. The value of a hit is 10 points. Upon completion of firing the CPQC, the scorer totals and signs the scorecard. Qualification standards are listed in the bottom right-hand corner of the record firing side of the scorecard. They are:

Expert	260 to 300
Sharpshooter	210 to 250
Marksman	160 to 200
Unqualified	below 160

NBC and night qualification is on a GO/NO-GO scoring system and recorded in the remarks column.

NBC: 7 target hits = GO

Night: 5 target hits = GO

c. Supply of Forms. DA Form 88 is available through normal publications supply channels (see Figure A-1).

C-7. TARGETS

Seven electric target device targets and E-type silhouettes for each firing lane are required. Aggressor figures may be superimposed on the silhouettes to add realism to the course of fire.

C-8. QUICK-FIRE TARGET TRAINING DEVICE

The QTTD may be procured locally. For durability and appearance, it should be made by the training aids section or an equally capable agency.

APPENDIX D

ALTERNATE REVOLVER QUALIFICATION COURSE

D-1. PROCEDURES

Once the soldier completes instructional firing, he must then fire the CPQC for record. If the CPQC is not available, then the soldier may fire the ARQC.

NOTE: The tower operator controls all reloading.

a. Procedures for Firing ARQC with the Caliber .38 Revolver.

(1) *Table 1:* Engage the 25-meter E-type silhouette target with rings from the standing position with six rounds of ammunition; given six rounds for the caliber .38 revolver on a 25-meter range during daylight. Within 21 seconds, from the standing position, engage the E-type silhouette target (see Figure B-1).

(2) *Table 2:* Engage the 25-meter target from the kneeling position with 12 rounds; given 12 rounds of ball ammunition and a caliber .38 revolver on a 25-meter range during daylight. Within 23 seconds, from a standing position, assume a good kneeling position and engage the target with six rounds. Repeat steps for next six rounds.

(3) *Table 3:* Engage the 25-meter target from the crouch position with 12 rounds; given 12 rounds of ball ammunition and a caliber .38 revolver on a 25-meter range during daylight. Within 23 seconds, from a standing position, assume a good crouch position and engage the target with the first six rounds. Repeat steps for next six rounds.

(4) *Table 4:* Engage the 25-meter target from the prone position with 10 rounds; given 10 rounds of ball ammunition and a caliber .38 revolver on a 25-meter range during daylight hours. Within 23 seconds, from a standing position, assume a good prone position, and engage the target with the first six rounds. Repeat steps for next four rounds, allowing only 18 seconds.

b. Firing Revolver Under Night Conditions. Engage the 25-meter target from a crouch position with 30 rounds; given 30 rounds of ball ammunition and a caliber .38 revolver on a 25-meter range during nighttime conditions. Within 60 seconds, engage six pop-up E-type silhouettes with six rounds. Reload only on command from the tower. Repeat steps for next 24 rounds.

c. Firing Revolver Under Simulated NBC Conditions. Engage the 25-meter target from a crouch position with 20 rounds; given 20 rounds of ball ammunition and a caliber .38 revolver on a 25-meter range under simulated NBC conditions during daylight. Within 40 seconds, engage the 25-meter pop-up target with six rounds. Reload only on command from the tower. Repeat steps for next 14 rounds.

NOTE: Night and NBC qualification is required IAW DA Pam 350-38.

D-2. CONDUCT OF FIRE

a. The following commands outline a step-by-step sequence for conducting range firing on the ARQC.

(1) *Table 1:* Standing position.

(a) The tower operator orders firers to move to the firing line and to prepare to fire. The caliber .38 rounds are issued to the scorer and given to the firer on command.

(b) The tower operator commands:

TABLE ONE, STANDING POSITION, SIX ROUNDS.
LOAD.
IS THE LINE READY?
THE FIRING LINE IS READY.
FIRERS, WATCH YOUR LANE!

(c) At the end of the prescribed firing time, the tower operator commands:

CEASE FIRE.

ARE THERE ANY ALIBIS?

(Allowable alibis are given two seconds for each round not fired.)

NOTE: For more information, see paragraph D-3.

UNLOAD AND CLEAR ALL WEAPONS.

IS THE FIRING LINE READY?

THE FIRING LINE IS NOW CLEAR.

FIRERS AND SCORERS, MOVE DOWNRANGE AND CHECK YOUR TARGETS.

(All weapons are cleared and left on table, or they are left at the firing line with the cylinder in the open position.)

FIRERS AND SCORERS, MOVE DOWNRANGE AND CHECK YOUR TARGETS.

(2) *Table 2:* Kneeling position.

(a) The tower operator orders firers to move to the firing line. Scorers are issued 12 rounds of caliber .38 ammunition to be given to the firer on command.

(b) The tower operator commands:

TABLE TWO, KNEELING POSITION, TWELVE ROUNDS.

FIRST SIX ROUNDS, TWENTY-THREE SECONDS.

RELOAD, SECOND SIX ROUNDS, TWENTY-THREE SECONDS.

LOAD FIRST SIX ROUNDS.

IS THE LINE READY?

THE FIRING LINE IS READY.

FIRERS, WATCH YOUR LANE!

(c) At the end of the prescribed firing time, the tower operator commands:

CEASE FIRE.
ARE THERE ANY ALIBIS?
UNLOAD AND CLEAR ALL WEAPONS.
LOAD SECOND SIX ROUNDS.
IS THE LINE READY?
THE FIRING LINE IS READY.
FIRERS, WATCH YOUR LANE!

(d) At the end of the prescribed firing time, the tower operator commands:

CEASE FIRE.
ARE THERE ANY ALIBIS?

NOTE: Allowable alibis are given two seconds for each round.

UNLOAD AND CLEAR ALL WEAPONS.
IS THE FIRING LINE CLEAR?
THE FIRING LINE IS NOW CLEAR.
FIRERS AND SCORERS, MOVE DOWNRANGE AND CHECK YOUR TARGETS.

NOTE: All weapons are cleared and left on a table, or they are left standing at the firing line with the cylinder in the open position. Then the firers and scorers move downrange to check their targets.

(3) *Table 3:* Crouch position.

(a) The tower operator orders the firers and scorers to move to the firing line. The scorers are issued 12 rounds of caliber .38 ammunition to be given to the firer on command.

(b) The tower operator commands:

TABLE THREE, CROUCH POSITION, TWELVE ROUNDS.
FIRST SIX ROUNDS, TWENTY-THREE SECONDS.
RELOAD, SECOND SIX ROUNDS, TWENTY-THREE SECONDS.

NOTE: All commands are the same as for Table 2.

(4) *Table 4:* Prone position.

(a) The tower operator orders the firers to move to the firing line. The firers are issued 10 rounds of caliber .38 ammunition.

(b) The tower operator orders:

TABLE FOUR, PRONE POSITION, TEN ROUNDS.

FIRST SIX ROUNDS, TWENTY-THREE SECONDS.

RELOAD, NEXT FOUR ROUNDS, EIGHTEEN SECONDS.

NOTE: All commands are the same as for Tables 1, 2, and 3. The scorers and firers replace all targets for the next firing order. Excess ammunition at the end of the course is turned in to the ammunition point.

b. The commands for the revolver night fire for record are as follows:

(1) The tower operator orders firers to move to the firing line. Scorers are issued 30 rounds to be given to the firer on command.

(2) The tower operator commands:

NIGHT FIRE, CROUCH POSITION, SIXTY SECONDS, SIX ROUNDS.

RELOAD ONLY ON COMMAND.

LOAD FIRST SIX ROUNDS.

IS THE FIRING LINE READY?

THE FIRING LINE IS READY.

FIRERS, WATCH YOUR LANE.

(3) At the end of the prescribed time, the tower operator commands:

CEASE FIRE.

ARE THERE ANY ALIBIS?

(Alibis are allowed 10 second for each round not fired.)

UNLOAD AND CLEAR ALL WEAPONS.

NOTE: These commands are repeated for each six rounds fired.

IS THE FIRING LINE CLEAR?

THE FIRING LINE IS CLEAR.

FIRERS AND SCORERS, MOVE DOWNRANGE AND CHECK YOUR TARGETS.

c. The commands for the revolver NBC fire for record are as follows:

(1) The tower operator orders firers to move to the firing line. Scorers are issued 20 rounds to be given to the firer on command.

(2) The tower operator commands:

GAS (Firers don protective masks.)

NBC FIRING, CROUCH POSITION, FORTY SECONDS, SIX ROUNDS.

RELOAD ONLY ON COMMAND.

LOAD FIRST SIX ROUNDS.

IS THE FIRING LINE READY?

THE FIRING LINE IS READY.

FIRERS, WATCH YOUR LANE.

(3) At the end of the prescribed time, the tower operator commands:

CEASE FIRE.

ARE THERE ANY ALIBIS?

(Alibis are allowed eight seconds for each round not fired.)

UNLOAD AND CLEAR ALL WEAPONS.

NOTE: These commands are repeated for each six rounds fired.

IS THE FIRING LINE CLEAR?

THE FIRING LINE IS NOW CLEAR.

(The tower operator also gives the command, ALL CLEAR.)

FIRERS AND SCORERS, MOVE DOWNRANGE AND CHECK YOUR TARGETS.

(All weapons are left on the firing line with cylinders open.)

NOTE: The scorers and firers replace all targets for the next firing order. Excess ammunition at the end of a table is turned in to the scorer and is not used by the firer in subsequent tables. At the completion of all four tables, ammunition is turned in to the ammunition point.

D-3. ALIBIS

If a malfunction of the weapon or the target occurs during firing, the scorer reports and records the malfunction. The firer is allowed one alibi at the completion of each table. All alibis are fired from the position in which the alibi occurred. Firing commands that apply are used to fire alibis.

D-4. SCORING

a. The firer is scored on the number of target hits during the prescribed time limit. He must achieve at least 24 hits and a score of 80 points to qualify. The target hits are then multiplied by the number inside the scoring rings to achieve a score. No credit is given for rounds fired after the command, CEASE FIRE. Shots that touch the next higher scoring ring are scored the next higher value (see Figure B-1.)

b. The qualification scores are:

Expert	160 to 200
Sharpshooter	120 to 159
Marksman	80 to 119

NBC and night firing are done on a GO/NO-GO scoring system and recorded in the remarks column.

NBC: 7 target hits = GO

Night: 5 target hits = GO

NOTE: For sample scorecard see Figure D-1.

c. Coaching is allowed during instructional firing but not during record fire. No one may assist the firer while he is taking position or after taking position at the firing point except for safety reasons.

ALTERNATE REVOLVER QUALIFICATION COURSE
For use of this form, see FM 23-35, the Proponet agency is TRADOC

NAME (Last, First, MI)				DATE
Tentpeg Joe M.				29MAR88
LANE NO.	**ORDER**	**UNIT**	**SSN**	
3	1	Co 2/29 Inf	333-33-3333	

TABLE 1 - STANDING POSITION: 6 Rounds - 21 Seconds

HITS X X X X X X — HITS 6
SCORE 5 5 4 4 5 3 — SCORE 26

TABLE 2 - KNEELING POSITION: 6 Rounds - 23 Seconds: Reload Under Control of the Tower
6 Rounds - 23 Seconds

HITS X X X X X X M X X X X X — HITS 11
SCORE 5 5 5 5 5 4 5 0 5 5 5 5 — SCORE 54

TABLE 3 - CROUCH POSITION: 6 Rounds - 23 Seconds: Reload Under Control of the Tower
6 Rounds - 23 Seconds

HITS X X X X X X X X X X X X — HITS 12
SCORE 5 5 5 5 5 5 5 5 3 5 5 5 — SCORE 58

TABLE 4 - PRONE POSITION: 6 Rounds - 23 Seconds; Reload Without Command
4 Rounds - 18 Seconds

HITS X X X X X X X X X M — HITS 9
SCORE 5 5 5 5 3 4 5 5 3 0 — SCORE 40

QUALIFICATION:
EXPERT 160-200
SHARPSHOOTER 120-159
MARKSMAN 80-119

TOTAL HITS 38
TOTAL SCORE 178

SCORER'S SIGNATURE	DATE	OFFICER'S SIGNATURE	DATE
[illegible]	29MAR88	[illegible]	29MAR88

REMARKS:
NIGHT Fire - GO
NBC Fire - GO

NOTES: 1. HITS ARE MARKED WITH "X." AND MISSES ARE MARKED WITH "M."

2. THE FIRER MUST ACHIEVE A MINIMUM OF 24 HITS AND A MINIMUM SCORE OF 80 TO QUALIFY.

DATA REQUIRED BY PRIVACY ACT OF 1974

AUTHORITY: 10USC3012g / Executive Order 9397. PRINCIPAL PURPOSE(S): Record individual's performance on Record Fire Range. ROUTINE USE(S): Evaluation of individual's proficiency and basis for determination of award of proficiency badge. SSN is used for positive identification purposes only. MANDATORY OR VOLUNTARY DISCLOSURE AND EFFECT ON INDIVIDUAL NOT PROVIDING INFORMATION: Voluntary. Individuals not providing information cannot be rated/scored on a mass basis.

DA Form 5705-R, SEP 88

FIGURE D-1. Example of completed Alternate Revolver Qualification Course form.

NOTE: See Appendix F for a blank copy of this form for local reproduction.

APPENDIX E

TRAINING SCHEDULES

To aid in the individual training phase, training schedules for the courses in pistol and revolver marksmanship training are described in this appendix. These schedules are based on the desirable number of training hours for a pistol or revolver course. They should be used as a guide in preparing lesson plans. Conditions may require a longer or shorter period to complete the training. When time is available, additional training should be included in the schedule. When suggested equipment and training aids are not available, the best that are available should be improvised or substituted. Each firer should be allowed 50 rounds for instructional firing and 40 rounds for record firing.

A. Pistol, Semiautomatic, Caliber 9-mm, Caliber .45 M1911A1; Revolver, Caliber .38 (Practice or Instructional Firing Course (12 Hours)

Period	Hours		Lesson	References	Training Facilities	Training Aids
	Peace	Mobilization				
			MECHANICAL TRAINING (4 HOURS)			
1	2	2	Characteristics, disassembly and assembly, functioning, and care and cleaning.	TM 9-1005-317-10, TM 9-1005-211-12, and TM 9-1005-226-14.	Clasroom or field.	For instructor: chalkboard, working model, projector and screen, cleaning equipment (for each man). For each group: Table or suitable ground cloth.
2	2	2	Malfunctions, stop-pages, immediate action, loading, unloading, ammunition, and safety precautions.	TM 9-1005-317-10, TM 9-1005-211-12, and TM 9-1005-226-14.	. . . do . . .	Same as period 1, plus ammunition display.
3	3	3	PREPARATORY MARKSMANSHIP TRAINING (16 HOURS) Coaching, aiming, grip, positions, trigger squeeze (to include double-action), target engagement, pencil triangulation exercise (M1911A1 only), slow-fire exercise.	Chapter 2 this manual.	. . . do . . .	For each man: One pistol with magazine, sheet of 1/8-inch bull's-eye, pencil with masking or cellophane tape. For all: E-silhouette.
4	2	2	RANGE FIRING (12 HOURS) Instructional firing Tables 1, 2, 3, 4, & 5, Combat Pistol Qualification Course.	App A this manual.	Live-fire range.	Equipment used in period 6 of the qualification course.

B. Pistol, Semiautomatic, Caliber 9-mm, Caliber .45 M1911A1; Revolver, Caliber .38 (Qualification Course) (12 Hours)

Period	Hours		Lesson	References	Training Facilities	Training Aids
	Peace	Mobilization				
			RANGE FIRING (4 HOURS)			
5	2	2	Instructional firing combat pistol qualification course, for practice with a coach or instructor.	Existing range regulations. App A this manual.	Pistol range.	For all: All equipment used for periods 3 and 4 plus scorecard and ammunition.
6	2	2	Record firing, Tables 1 through 5, combat pistol qualification course.	App A this manual.	. . . do . .	. . . . . . do . . . . .

B. Pistol, Semiautomatic, Caliber 9-mm, Caliber .45 M1911A1; Revolver, Caliber .38 (Qualification Course) (12 Hours) (Continued)

Period	Hours		Lesson	References	Training Facilities	Training Aids
	Peace	Mobilization				
			MECHANICAL TRAINING (4 HOURS)			
1	2	2	Characteristics, disassembly and assembly, functioning, and care and cleaning.	TM 9-1005-317-10, TM 9-1005-211-12, and TM 9-1005-226-14.	Clasroom or field.	For instructor: chalkboard, working model, projector and screen, cleaning equipment (for each man). For each group: Table or suitable ground cloth.
2	2	2	Malfunctions, stop-pages, immediate action, loading, unloading, ammunition, and safety precautions.	TM 9-1005-317-10, TM 9-1005-211-12, and TM 9-1005-226-14.	. . . do . . .	Same as period 1, plus ammunition display.
3	2	2	PREPARATORY MARKSMANSHIP TRAINING (16 HOURS) Coaching, aiming, grip, positions, trigger squeeze (to include double-action), target engagement, pencil triangulation exercise (M1911A1 only), slow-fire exercise.	Chapter 2 this manual.	. . . do . . .	For each man: One pistol with magazine, sheet of 1/8-inch bull's-eye, pencil with masking or cellophane tape. For all: E-silhouette.
4	2	2	Review and examination.	All previous references.	. . . do . .	For all: All equipment used in previous periods.

APPENDIX F

REPRODUCIBLE FORMS

ALTERNATE PISTOL QUALIFICATION COURSE

For use of this form, see FM 23-35, the Proponet agency is TRADOC.

NAME: (Last, First, MI)	DATE

LANE NO.	ORDER	UNIT	SSN

TABLE 1 - STANDING POSITION: 1 Magazine - 7 Rounds - 21 Seconds

HITS ☐☐☐☐☐☐☐ HITS ____

SCORE ☐☐☐☐☐☐☐ SCORE ____

TABLE 2 - KNEELING POSITION: First Magazine - 6 Rounds - 45 Seconds
Second Magazine - 7 Rounds

HITS ☐☐☐☐☐☐☐☐☐☐☐☐☐ HITS ____

SCORE ☐☐☐☐☐☐☐☐☐☐☐☐☐ SCORE ____

TABLE 3 - CROUCH POSITION: 2 Magazines - 5 Rounds Each - 35 Seconds

HITS ☐☐☐☐☐☐☐☐☐☐ HITS ____

SCORE ☐☐☐☐☐☐☐☐☐☐ SCORE ____

TABLE 4 - PRONE POSITION: 2 Magazines - 5 Rounds Each - 35 Seconds

☐☐☐☐☐☐☐☐☐☐ HITS ____

☐☐☐☐☐☐☐☐☐☐ SCORE ____

QUALIFICATION:

EXPERT 160-200
SHARPSHOOTER 120-159
MARKSMAN 80-119

TOTAL HITS ____

TOTAL SCORE ____

SCORER'S SIGNATURE	DATE	OFFICER'S SIGNATURE	DATE

REMARKS:

NOTES: 1. HITS ARE MARKED WITH "X," AND MISSES ARE MARKED WITH "M."

2. THE FIRER MUST ACHIEVE A MINIMUM OF 24 HITS AND A MINIMUM SCORE OF 80 POINTS TO QUALIFY.

DATA REQUIRED BY PRIVACY ACT OF 1974

AUTHORITY: 10USC3012g/Executive Order 9397. PRINCIPAL PURPOSE(S): Records individual's performance on Record Fire Range. ROUTINE USE(S): Evaluation of individual's proficiency and basis for determination of award of proficiency badge. SSN is used for positive identification purposes only. MANDATORY OR VOLUNTARY DISCLOSURE AND EFFECT ON INDIVIDUAL NOT PROVIDING INFORMATION: Voluntary. Individuals not providing information cannot be rated/scored on a mass basis.

DA Form 5704-R, SEP 88

FIGURE F-1. DA Form 5704-R, Alternate Pistol Qualification Course.

ALTERNATE REVOLVER QUALIFICATION COURSE

For use of this form, see FM 23-35; the Proponet agency is TRADOC

NAME (Last, First, MI)			DATE
LANE NO.	ORDER	UNIT	SSN

TABLE 1 - STANDING POSITION: 6 Rounds - 21 Seconds

HITS / SCORE

HITS ____
SCORE ____

TABLE 2 - KNEELING POSITION: 6 Rounds - 23 Seconds; Reload Under Control of the Tower
6 Rounds - 23 Seconds

HITS / SCORE

HITS ____
SCORE ____

TABLE 3 - CROUCH POSITION: 6 Rounds - 23 Seconds; Reload Under Control of the Tower
6 Rounds - 23 Seconds

HITS / SCORE

HITS ____
SCORE ____

TABLE 4 - PRONE POSITION: 6 Rounds - 23 Seconds; Reload Without Command
4 Rounds - 18 Seconds

HITS / SCORE

HITS ____
SCORE ____

QUALIFICATION:
EXPERT 160-200
SHARPSHOOTER 120-159
MARKSMAN 80-119

TOTAL HITS ____
TOTAL SCORE ____

SCORER'S SIGNATURE	DATE	OFFICER'S SIGNATURE	DATE

REMARKS:

NOTES: 1. HITS ARE MARKED WITH "X," AND MISSES ARE MARKED WITH "M."

2. THE FIRER MUST ACHIEVE A MINIMUM OF 24 HITS AND A MINIMUM SCORE OF 80 TO QUALIFY.

DATA REQUIRED BY PRIVACY ACT OF 1974

AUTHORITY: 10USC30129 / Executive Order 9397. PRINCIPAL PURPOSE(S): Record individual's performance on Record Fire Range. ROUTINE USE(S): Evaluation of individual's proficiency and basis for determination of award of proficiency badge. SSN is used for positive identification purposes only. MANDATORY OR VOLUNTARY DISCLOSURE AND EFFECT ON INDIVIDUAL NOT PROVIDING INFORMATION: Voluntary. Individuals not providing information cannot be rated/scored on a mass basis.

DA Form 5705-R, SEP 88

FIGURE F-2. DA Form 5705-R, Alternate Revolver Qualification Course.

GLOSSARY

APQC	alternate pistol qualification course
ARQC	alternate revolver qualification course
AR	Army regulation
CID	criminal investigations division
CPQC	combat pistol qualification course
CTAA	common table of allowances
DA	Department of the Army
FM	field manual
HQ	headquarters
mm	millimeter
NATO	North Atlantic Treaty Organization
NBC	nuclear, biological, chemical
NG	Army National Guard
OIC	officer in charge
QTTD	quick-fire target training device
RQC	revolver qualification course
TM	technical manual
TRADOC	US Army Training and Doctrine Command
USAR	United States Army Reserve

REFERENCES

REQUIRED PUBLICATIONS

Required publications are sources that users must read in order to understand or to comply with this publication.

Army Regulation (AR)

385-63	Policies and Procedures for Firing Ammunition for Training Target Practice, and Combat.

Department of the Army Forms (DA Forms)

88	Combat Pistol Qualification Course Scorecard
5704-R	Alternate Pistol Qualification Course
5705-R	Alternate Revolver Qualification Course

Technical Manuals (TMs)

9-1005-206-14&P1	Operator's, Organizational, Direct Support and General Support Maintenance Manual Including Repair Parts and Special Tools List for Revolver, Caliber .38 Special: Smith and Wesson, Military and Police, M10, Round Butt, 4-Inch Barrel, 2-Inch Barrel; Square Butt, 4-Inch Barrel and Revolver, Caliber .38 Special: Ruger Service Six, 4-Inch Barrel, M108; Square Butt w/o Lanyard Loop, w/Lanyard and Round Butt w/Lanyard Loop.
9-1005-211-12	Operator and Organizational Maintenance Manual (Including Basic Issue Items List and

	Repair Parts and Special Tools List): Pistol, Caliber .45, Automatic, M1911A1, with Holster, Hip (1005-673-7965); with Holster, Shoulder (1005-561-2003).
9-1005-226-14	Operation and Unit Maintenance: Caliber .22 High Standard Automatic Pistol (Supermatic) Caliber .22 Ruger Mark I Automatic Pistol (Target Model) (6 7/8-Inch Barrel); Caliber .38 Special Smith and Wesson Revolver (Masterpiece); Caliber .30-06 Winchester Rifle, Model 70 (Special Match Grade; Caliber .22 Winchester Rifle, Model 52; Caliber .22 Remington Rifle, 40x51 (National Match) and Front and Rear Sights.
9-1005-317-10	Operator's Manual for Pistol, Semiautomatic, 9-mm, M9 (1005-01-118-2640).
9-1300-200	Ammunition, General.

RELATED PUBLICATIONS

Related publications are sources of additional information. They are not required in order to understand this publication.

Army Regulations (ARs)

140-1	Marksmanship Training and Competitive Program.
350-6	Army-Wide Small Arms Competitive Marksmanship.
920-30	Rules and Regulations for National Matches and other Excellence-in-Competitive (EIC) Matches.

Common Tables of Allowances (CTAs)

8-100	Army Medical Department Expendable/ Durable Items.
50-970	Expendable/Durable Items (Except: Medical, Class V, Repair Parts and Heraldic Items).

Department of the Army Pamphlet (DA Pam)

350-38	Standards in Weapons Training
738-750	The Army Maintenance Management System (TAMMS).

Field Manuals (FMs)

3-4	NBC Protection.
3-5	NBC Decontamination.
21-11	First Aid for Soldiers.
21-75	Combat Skills of the Soldier.
25-7	Training Ranges.

Technical Manuals (TMs)

9-1005-211-35	Direct Support, General Support, and Depot Maintenance Manual (Including Repair Parts and Special Tools List): Pistol, Caliber .45, M1911A, with Holster (1005-673-7965) and M1911A1 (1005-561-2003).
9-1005-317-23&P	Unit and Intermediate Direct Support Maintenance Manual Including Repair Parts and Special Tools List for Pistol, Semiautomatic, 9-mm, M9 (1005-01-118-2640).
9-1300-206	Ammunition and Explosives Standards.
9-6920-210-14	Operator's, Organizational, Direct Support and General Support Maintenance Manual (Including Basic Issue Items List and Repair Parts List) for Small Arms Targets and Target Material.
43-0001-27	Army Ammunition Data Sheets: Small Caliber Ammunition.

INDEX

A

aiming, 11, 12

air-operated pistol, .177-mm, 35, 36

alibis, 52, 60, 70, 79

  alternate pistol qualification course (APQC), 45, 55–62

  alibis, 60

  conduct of fire, 56

  form, 62, 86

  scoring, 60

  tables, 55, 59

alternate revolver qualification course (ARQC), 45, 63, 73–79

  alibis, 79

  conduct of fire, 74

  form, 80, 87

  scoring, 79

  tables, 73, 74

B

ball-and-dummy method, 32

basic marksmanship, 7–16

  aiming, 11, 12

  breath control, 13

  fundamentals, 7

  phases of training, 7

  positions, 16

target engagement, 15
trigger squeeze, 14
breath control, 13

C

calling the shot, 32
coaching, 31
combat marksmanship, 22–31
nuclear, biological, chemical (NBC) firing, 31
poor visibility firing, 30
reloading techniques, 28
target engagement, 23
techniques of firing, 22
traversing, 23, 24, 25, 26, 27
combat pistol qualification course (CPQC), 45–54, 46, 55, 63
alibis, 52
conduct of fire, 48
rules, 52
scorecard, 54, 53
tables, 45, 46
targets, 53
combat reloading techniques, 28
one-handed, 29
rapid, 29
tactical, 29

E

equipment data,
pistol, semiautomatic, .45 caliber, M1911 and M1911A1, 2
pistol, semiautomatic, 9 mm, M9, 1
revolver, caliber .38, 4
Ruger revolver, caliber .38, 4
Smith and Wesson revolver, caliber .38, 4

F

flash sight picture, 22
forms, reproducible, 86, 87

G

grip, 7
  isometric tension, 11
  one-hand, 7, 8
  point of aim, 11
  two-hand, 8
    fist, 9
    palm-supported, 9, 10
    Weaver, 10

H

hand-and-eye coordination, 22

I

instructional practice and record qualification firing
  safety, 43

M

marksmanship training
  basic marksmanship, 7–16
  coaching and training aids, 31–40
  combat marksmanship, 22–31
  safety, 42–43

N

nuclear, biological, chemical (NBC) firing, 31, 48, 51, 53, 56, 65, 72, 74

O

operation
    pistol, automatic, .45 caliber, M1911 and M1911A1, 3
    pistol, semiautomatic, 9-mm, M9, 2

P

pencil triangulation exercise, 33
pistol, automatic, .45 caliber, M1911 and M1911A1
    equipment data, 2
    operation, 3
pistol, semiautomatic, .45 caliber, M1911 and M1911A1, 2
pistol, semiautomatic, 9 mm, M9, 1
    equipment data, 1
    operation, 2
point of aim, 11
poor visibility firing, 30
positions, 16
    crouch, 19
    kneeling, 17, 18
    kneeling supported, 21
    pistol-ready, 16, 17
    prone, 19, 20
    standing without support, 17, 18
    standing with support, 20

Q

quick-fire point shooting, 23
quick-fire sighting, 23
quick-fire target training device (QTTD), 36, 38, 54
    dimensions, 37

R

range firing courses, 7, 40
    alternate pistol qualification course (APQC), 41
    alternate revolver qualification course (ARQC), 41
    combat pistol qualification course (CPQC), 41

revolver qualification course (RQC), 41
revolver, caliber .38, 4
    equipment data, 4
    operation, 5
revolver qualification course (RQC), 63–72
    alibis, 70
    conduct of fire, 65
    rules, 71
    scorecard, 72
    tables, 63, 64
    targets, 72
Ruger revolver, caliber .38, 4
    equipment data, 4
    operation, 5

S

safety, 42–44
    after firing, 43
    before firing, 42
    during firing, 43
    instructional practice and record qualification firing, 43
    requirements, 42
scorecard, DA Form 88, 54, 53, 62, 72
slow-fire exercise, 35
Smith and Wesson revolver, caliber .38, 4
    equipment data, 4
    operation, 5

T

target engagement, 15, 23
targets, 36, 37, 38, 62
techniques of firing
    flash sight picture, 22
    hand-and-eye coordination, 22
    quick-fire point shooting, 23
    quick-fire sighting, 23

training aids
    air-operated pistol, .177-mm, 35, 36
    quick-fire target training device (QDDT), 36
training schedules, 82–83
traversing, 23, 24, 25
    kneeling 360°, 23, 26
    training method, 28
trigger squeeze, 14